THE FOSSILS OF
FLORISSANT

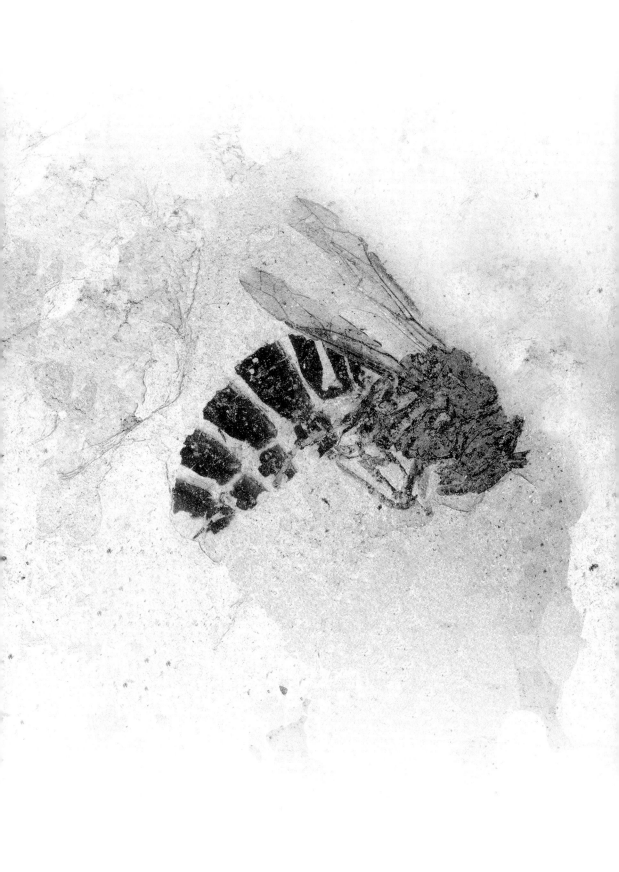

THE FOSSILS OF

FLORISSANT

HERBERT W. MEYER

SMITHSONIAN BOOKS • Washington and London

© 2003 by the Smithsonian Institution
All rights reserved

Copy editor: Natasha Atkins
Production editor: Duke Johns
Designer: Janice Wheeler

Library of Congress Cataloging-in-Publication Data
Meyer, Herbert W. (Herbert William), 1954–
　The fossils of Florissant / Herbert W. Meyer
　　p.　cm.
　Includes bibliographical references and index.
　　ISBN 1-58834-107-0 (alk. paper).
　1. Fossils—Colorado—Florissant region—Classification.　2. Florissant Fossil Beds National Monument (Colo.)　3. Florissant Fossil Beds National Monument (Colo.)—Bibliography.　I. TItle
QE747.C6 M49 2003
560′.1784′09788—dc21　2002030435

British Library Cataloguing-in-Publication Data available

Manufactured in China, not at government expense
10　09　08　07　06　05　04　03
5　4　3　2　1

♾ The paper used in this publication meets the minimum requirements of the American National Standard for Information Sciences—Permanence of Paper for Printed Library Materials ANSI Z39.48-1984.

For permission to reproduce illustrations appearing in this book, please correspond directly with the owners of the works, as listed in the individual captions. Uncredited photographs belong to the author. Smithsonian Books does not retain reproduction rights for these illustrations individually, or maintain a file of addresses for photo sources.

DEDICATED TO THE MANY PALEONTOLOGISTS

WHOSE WORK OVER A PERIOD OF 130 YEARS

HAS CREATED THE KNOWLEDGE

THAT MAKES THE WONDERS OF FLORISSANT

KNOWN TO THE WORLD.

That creatures so minute and fragile as insects, creatures which can so feebly withstand the changing seasons as to live, so to speak, but a moment, are to be found fossil, engraved, as it were, upon the rocks or embedded in their hard mass, will never cease to be a surprise to those unfamiliar with the fact.

—Samuel H. Scudder, from his 1890 introduction to
The Tertiary Insects of North America

We get all tangled up with the present. The present is just a little flick in time between the past and the future. Things keep going on and on. We are just in this particular little time interval, and it seems so important to us.

—Harry D. MacGinitie, from a 1979 interview at Florissant

CONTENTS

ix   PREFACE
xi   ACKNOWLEDGMENTS

1   INTRODUCTION
5   HISTORY OF RESEARCH AND CONSERVATION
21   GEOLOGIC SETTING AND PROCESSES OF FOSSILIZATION
45   RECONSTRUCTING THE ANCIENT ECOSYSTEM
73   FOSSIL PLANTS
127   FOSSIL SPIDERS, INSECTS, AND OTHER INVERTEBRATES
175   FOSSIL VERTEBRATES
187   EPILOGUE

189   APPENDIX 1. A COMPLETE LISTING OF THE FOSSIL ORGANISMS FROM FLORISSANT
233   APPENDIX 2. MUSEUMS WITH SIGNIFICANT FLORISSANT COLLECTIONS
237   GENERAL REFERENCES
239   BIBLIOGRAPHY
255   INDEX

PREFACE

This book brings together, for the first time, the story of all the fossilized species that have been described from the Florissant fossil beds in the Rocky Mountains of Colorado. Now part of Florissant Fossil Beds National Monument, this site has been studied for more than 130 years, resulting in the description of hundreds of species in about 400 scientific publications. Indeed, more than a dozen museums worldwide now hold Florissant type specimens—those important specimens that are set aside to define the characteristics of new species.

Until recently, most of the fossils from Florissant were so widely dispersed both in the literature and between different museums that it was a challenge for paleontologists to locate them. No complete list of all of the described species, especially of the insects, had ever been compiled. I began work in 1995 to compile a new database that documents Florissant specimens both in publications and in museums. This effort has involved numerous visits to all of the museums that keep the most important Florissant collections—from Berkeley, California, to Washington, D.C., to London, England. In the end, all of the specimens discussed in the scientific literature were examined, recorded, photographed (many for the first time), and placed into a database maintained by the National Park Service. This database is available as a website by link from the website for Florissant Fossil Beds National Monument at http://www.nps.gov/flfo, or (as of this writing) go directly to http://www.planning.nps.gov/flfo/. It includes photographic images for almost all of the published Florissant fossils as well as digital copies of many older scientific publications about Florissant.

Despite the fame of Florissant, and even though there are at least 40,000 specimens at the various museums, only a small fraction of Florissant fossils are on public exhibit around the world. This book now makes it possible to illustrate some of the finest fossils from Florissant—fossils that are otherwise inaccessible to the public—and to summarize the significance of this important site in the world of paleontology. All of the photographs of fossils illustrated in this book are from Kodachrome transparencies that I took, unless otherwise noted.

The bibliography at the end of the book lists virtually all of the publications that pertain to Florissant, and the general references section cites sources of background information. A section at the end of each chapter lists a selection of relevant references on the topics discussed. The appendixes provide a concise compendium for all of the many described species of ancient organisms in the fossil record at Florissant, and the museums that house the most important collections. It is my hope that this book will arouse the interest of paleontologists and amateurs alike, stimulating new research and providing a better understanding of life in the earth's past.

ACKNOWLEDGMENTS

The U.S. National Park Service provided the financial support that made possible the completion of the Florissant database and this book. Many of my interns and seasonal staff in paleontology at Florissant Fossil Beds National Monument during the past seven years have contributed to projects that helped make this book possible: Amanda Cook, Jess DeBusk, Michelle Dooley, Melissa Hicks, Tobin Hieronymus, Scotty Hudson, Trudy Kernan, April Kinchloe, Cayce Lillesve, Rebecca Lincoln, Beth Simmons, Ta-Shana Taylor, and Matt Wasson. Their combined efforts, especially those of Amanda Cook, April Kinchloe, and Matt Wasson, have involved thousands of hours of computer time in helping to create the database of Florissant fossils upon which much of this book is based. Superintendent Jean Rodeck gave me with the unrestricted freedom and flexibility to pursue the idea of writing this book. Linda Lutz-Ryan generously produced the original artwork illustrated in several of the figures, and she drafted the diagram for Figure 39. Michelle Dooley provided assistance in organizing the figures. Tacy Smout assisted in formatting the bibliography and appendix, and she drafted the geologic time chart. John Fraser completed about half of the collection inventory at Harvard University, and Jim McChristal compiled historical information that was useful in completing part of the chapter on Florissant's history.

Boyce Drummond undertook the arduous task of reevaluating the higher taxonomic classification (above the rank of genus) for the insects, and completed much of the work that went into tabulating that part of the taxonomic list in Appendix 1. This compilation was assisted by Amanda Cook and Matt Wasson.

All of the museums that house the Florissant collections have been very generous and cooperative in providing access to the fossils of Florissant, and in granting permission to illustrate some of their finest specimens. I am thankful to the staffs of these museums for accommodating my visits, providing working space and access to their collections, and helping to resolve various cataloging problems. These institutions and individuals are as follows: National Museum of Natural History (Scott Wing, Bill DiMichele, Conrad Labandeira, Jann Thompson, Mark Florence, and Robert Purdy); Harvard Museum of Comparative Zoology (Phil Perkins, Brian Farrell, Laura Leibensperger, and Raymond Paynter); American Museum of Natural History (Bushra Hussaini, Ivy Rutzky, Neil Landman, David Grimaldi, and Malcolm McKenna); Yale Peabody Museum of Natural History (Leo Hickey, Linda Klise, Tim White, and Mary Ann Turner); University of California Museum of Paleontology (Diane Erwin and Howard Schorn); University of Colorado Museum (Peter Robinson and Paul Murphey); Denver Museum of Nature & Science (Kirk Johnson and Logan Ivy); Colorado School of Mines (Ginny Mast); Carnegie Museum of Natural History (Albert Kollar, Ilona Wyers, and Elizabeth Hill); Paul R. Stewart Museum of Waynesburg College (James Randolph); Field Museum of Natural History (Jenny McElwain, Peter Wagner, and Lance Grande); Milwaukee Public Museum (Paul Mayer and Peter Sheehan); Florida Museum of Natural History (Steve Manchester and Roger Portell); San Diego Natural History Museum (Tom Deméré); California Academy of Sciences; The Natural History Museum, London (Andrew Ross, Tiffany Foster, Peter Forey, and Cedric Shute); National Museums of Scotland (Liz Hide); and University of Glasgow Hunterian Museum (Neil Clark).

Draft versions of various sections of the text were reviewed by several people, among them some of the most qualified experts on Florissant's geology and paleontology. I appreciate the reviews provided by Boyce Drummond, Emmett Evanoff, Steve Manchester, Greg McDonald, and Neal O'Brien. Formal reviews of the entire manuscript were provided by Derek Briggs, Leo Hickey, and Donald Prothero, all of whom contributed valuable constructive criticism. Other reviewers for portions of the manuscript were Amanda Cook, Rebecca Lincoln, and Jean Rodeck. Emmett Evanoff also provided commentary that was valuable in developing the artistic reconstructions of the area's geology. Information and ideas about specific topics were provided by Bob Chandler, Ted Fremd, Laura Leibensperger, Steve Manchester, Jenny McElwain, Doug Nichols, Neil O'Brien, Howard Schorn, Dena Smith, Eugene Stoermer, and Elisabeth Wheeler. Vincent Burke, Nicole Sloan, Duke Johns, and Janice

Wheeler of Smithsonian Books guided me through the publication process, and Natasha Atkins contributed many improvements to the book during copy editing.

Finally, I want to recognize the late Harry D. MacGinitie, who shared with me his wisdom of paleobotany during his years of active retirement. Mac knew the fossil plants from Florissant better than anyone else ever has, and he was always an inspirational influence during my eleven years at Berkeley.

Many paleontologists and geologists of the past thirteen decades have contributed significant portions of their careers to studying, describing, and interpreting the fossil deposits of Florissant. It is their work that has created the body of knowledge reported in this book. It has not been practical to cite the contributor for every bit of information summarized in the text, but the bibliography cites their works. To these paleontologists, collectively, we owe our understanding of the Florissant fossil beds.

INTRODUCTION

Left: Detail of fruit of the golden-rain tree, *Koelreuteria alleni.* Courtesy of the University of Colorado Museum.

Figure 1. The Florissant fossil beds are located at an elevation of 2,500–2,600 meters in the Rocky Mountains of Colorado. This view looks east across Florissant Fossil Beds National Monument, with Pikes Peak on the horizon. File photograph, Florissant Fossil Beds National Monument.

High in the Rocky Mountains of Colorado, in the shadow of Pikes Peak, lies one of the world's richest and most diverse records of past life (Figures 1–3). Like an ancient time capsule, these fossils from a lake deposit at Florissant give scientists and visitors a glimpse of a long-lost community that inhabited this region during the latest part of the Eocene Epoch 34 million years ago. During this time, an active volcano similar to modern Mount Saint Helens towered above the Florissant landscape. Mudflows from this volcano buried a forest of redwood trees and later caused the lake to form. Florissant's largest and most conspicuous fossils are the stumps of these giant petrified trees. Most of the fossils, however, are leaves, fruits, and insects, delicately preserved in fragile, paper-thin lake shale. With about 1,700 described fossil species, Florissant ranks among the world's most diverse paleontological treasures.

Figure 2. The modern valley at Florissant preserves 34-million-year-old lake and volcanic deposits of the late Eocene Epoch. Today's Rocky Mountain Montane Conifer Forest lives in a very cold environment here, contrasting sharply with the warmer world in which Florissant's fossils lived. File photograph, Florissant Fossil Beds National Monument.

Figure 3. *(right)* Map showing location of Florissant, Colorado.

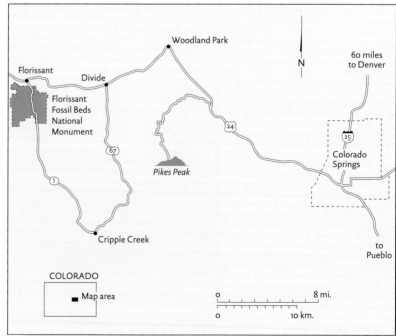

Scientific interest in the Florissant fossils began in the 1870s. At that time and during the following decades, large collections were amassed and removed for study by researchers from many universities and museums. Most of our knowledge about the fossil organisms from Florissant has come from the study of these early collections, which are now stored in museums from California to London. Although the fossil plants were carefully evaluated and revised by paleobotanist Harry MacGinitie in 1953, study of the fossil insects has progressed only slowly since the earliest work more than a century ago. Only recently has all of the information about these fossils been assembled into a single database, making it easily accessible in one place.

The significance of the Florissant fossils was evident even to the earliest paleontologists who worked there. The site was one of the most noteworthy of any to be recognized by the earliest scientific surveys of the American West. Numerous studies of these fossils over the years have demonstrated why Florissant is so important in the world of paleontology. In many regards, Florissant stands as a top contender among the world's fossil sites in claims for "the most," "the biggest," and "the only." Species diversity alone attests to Florissant's significance. The petrified forest includes tree stumps that are among the world's largest in circumference. No other fossil of a petrified redwood trio, or a tsetse fly, is known to exist. More important, however, is the evidence that Florissant provides about climate change, the evolution of plant and animal communities, the timing of mountain uplift in the Rocky Mountains, and the mechanisms of exceptional fossil preservation. Florissant preserves the best example of an upland ecosystem that existed immediately preceding a very significant cooling of the world's climate. The fossil plants reveal the vital clues for unraveling the nature of this climate change, and also for determining the elevation of the southern Rocky Mountains 34 million years ago. The ancient Florissant ecosystem was unique, and its plants and insects hold important evidence about the evolution of North American biotic communities and their response to this important time of change.

RELATED REFERENCES
Overview of Florissant's significance: Cockerell (1908y); Meyer and Weber (1995)
Database of Florissant fossils: Meyer (1998); Meyer et al. (2002)

HISTORY OF RESEARCH AND CONSERVATION

Our examination of Florissant's fossils begins not in the Eocene 34 million years ago, but with the beginning of human observations at Florissant. What the Utes—the Native Americans who lived in this region—thought of Florissant and its mysterious stone trees and their relation to creation legends remains a mystery. One might imagine that massive petrified plants evoked some interest, but we are left only with conjecture. We do know that the site made an impression on the early settlers of Colorado. By the 1860s, Colorado was becoming increasingly populated by newcomers and they certainly took note of the Florissant fossil beds.

THE EARLY YEARS OF DISCOVERY AND EXPLOITATION

One early immigrant who had a large impact on the region's settlement was Judge James Castello, who came from Florissant, Missouri, a town that took its name from the French word for flowering. Castello came to the area to build a general store and hotel, soon followed by a post office in 1872 that established the name of the town as Florissant. At last, it was a place on the map.

Early newspaper reports in the 1860s and 1870s told of Florissant's fossil treasures. The *Daily Central City Register* reported in 1871 that Florissant had "a petrified forest near which are found, between sedimentary layers, most beautiful imprints of leaves differing entirely from any that grow in the valley now-a-days." At first the area was too remote to attract more than the hardiest tourist, but that changed with the coming of the Colorado Midland Railway to Florissant in 1887. One train, the "Wildflower Excursion," brought droves of

Left: Detail of the compound leaf of *Rhus stellariaefolia*. Courtesy of The Natural History Museum, London.

Figure 4. The Colorado Midland Railway reached Florissant in 1887, making the area much more accessible to tourists. In the early 1900s, special excursions brought passengers to the high mountain meadows and allowed them to stop and collect fossils from this site near the town of Florissant. This photograph was taken from the rear of a train on 14 August 1902. Courtesy of the Colorado Historical Society; Buckwalter Book I, no. 63.

tourists to Florissant to see the summer wildflowers, visit the petrified forest, and collect fossil leaves and insects from a cut alongside the tracks (Figure 4).

Protection of the area was not to happen for nearly a century, and in the meantime, souvenir-hunting visitors did not hesitate to remove petrified wood. By the 1890s, much of the original petrified forest had been carted off. *The Creede Candle* reported in 1893 that "the oldest settlers thereabouts remember that 20 years ago there were 20 of these petrified trunks standing erect beside numerous petrified logs lying over the ground. All have been removed by tourists and relict hunters until now one of the greatest and rarest natural curiosities of the world has been despoiled. The trunks and logs have been sawed up and broken to pieces and taken East." Other reports indicate that horizontal trees with attached limbs once lay on the ground. One of the early paleontologists, Samuel Scudder, told of petrified trees that were originally five or six meters high, but "piecemeal they have been destroyed by vandal tourists, until now not one of them rises more than a meter above the surface of the ground, and many of them are entirely leveled." There was even interest in transporting a stump back east for the United States Centennial Exhibition in 1876 and the World's Columbian Exposition in 1893. An attraction known

HISTORY OF RESEARCH AND CONSERVATION 7

as the Big Stump was one of the huge stumps that survived these early exploitations, but not because people hadn't tried. A wooden framework was built around it, and saws were used in an attempt to cut this giant (weighing as much as 60 metric tons!) into manageable pieces that could be loaded onto a train and shipped east. Remnants of the broken saw blades, deeply embedded in the stump and still visible today, attest to the ultimate failure of this futile effort (Figure 5).

THIRTEEN DECADES OF SCIENTIFIC DISCOVERY

During the 1860s and 1870s, the geology of the American West was explored for the first time by several government-sponsored scientific surveys. One of these was the Hayden Survey of the Great Plains and Rocky Mountains, which included the exploration and mapping of the Colorado mountains. In 1873, Hayden Survey geologist A. C. Peale visited the Florissant area. Peale was the first to describe the geology of the "South Park District," which included Florissant, and in his report for the U.S. Geological and Geographical Survey of the Territories in 1874, he noted that ". . . around the settlement of Florissant is an irregular basin filled with modern [sic] lake deposits. The entire basin is not more than five miles in diameter. The deposits extend up the branches of the creek, which all unite near Florissant. Between the branches are granite islands appearing above the beds, which themselves rest on granite." Peale noted the "clay shales with fossil leaves" as well as a locality with 20 or 30 silicified stumps.

Figure 5. A deep cut and the remnant of a broken saw blade still remain in the Big Stump well over a century after attempts were made to cut it into pieces and transport it back east for exhibit. Even in several pieces, this stump—estimated to weigh more than 60 metric tons—would have been almost unmanageable to move in those days.

Paleontologists, too, were attracted by the unique fossils. Beginning in the 1870s, they came to Florissant to collect fossils for scientific research. Theodore Mead, a young college student who visited the area in 1871, made the first collections to end up in the hands of a paleontologist. When Samuel Scudder saw what Mead had found, his interests were aroused. Already emerging as America's foremost paleoentomologist, Scudder (Figure 6) was about to embark on a project at Florissant that helped to crown his distinguished career.

Scudder studied under the great naturalist Louis Agassiz at Harvard and had a long-term affiliation with the Boston Society of Natural History. He collected

Figure 6. Samuel H. Scudder (1837–1911) was one of the first paleontologists to visit Florissant and publish on its fossils. Educated at Harvard, he worked with the Hayden Survey to describe many new species of fossil insects from Florissant. In this portrait, he holds a small fossil insect in his hand. Photograph courtesy of the Ernst Mayr Library, Museum of Comparative Zoology, Harvard University.

at Florissant for five days during the summer of 1877, and returned at least in 1881 or 1882 (Figure 7) and again in 1889. During the first visit, he was accompanied by geologist Arthur Lakes, who produced the first geologic map of the Florissant area (Figure 8). The two men were hosted by Judge Castello, founder of Florissant, during their visit. Scudder collected from a site near the Big Stump, where he measured the first detailed stratigraphic section of the lake deposits. From 1886 until 1892, he held a position as paleontologist for the U.S. Geological Survey and worked with the Hayden Survey to study the new fossil insects from the West.

Figure 7. This historic photograph documents the site at which Scudder collected and measured the stratigraphic section published in his 1890 monograph. An accompanying handwritten note indicates: "I send you herewith a photograph taken by my boy (then 12 years old) last year in Colorado. It represents the hill at Florissant where most of the insect specimens were obtained. The right hand end is where my section was made and where your Princeton party worked. Yours sincerely, Samuel H. Scudder." The Big Stump (not visible in the photograph) is located at the base of the left side of the ridge. Scudder's only son was born in September 1869, and hence the photograph must have been taken in 1881 or 1882. The photograph provides rare and valuable documentation of one of the earliest collecting sites. Photograph from the collection of Florissant Fossil Beds National Monument.

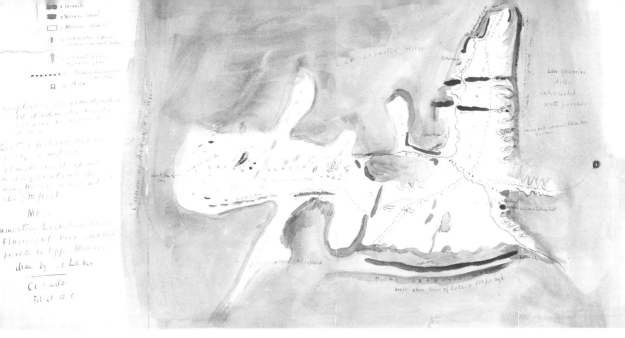

Figure 8. The geology of the Florissant area was first mapped by Arthur Lakes, who visited Florissant with Samuel Scudder in 1877 and produced this watercolor map in 1878. The legend describes it as "Map of Sedimentary Lacustrine basin at Florissant near South Park, Supposed to be Upper Miocene, drawn by A. Lakes. Colorado, Feb 20, 1878." The blue shaded areas were described as granite, the brown areas as volcanic lavas, and the uncolored white areas as Miocene sediment. A few cultural features, such as "Judge Castello's house," are also shown. The map is reasonably accurate in depicting the general distribution of the Florissant Formation as we understand it today (compare with the geologic map in Figure 25). Orientation is with north to the right. The original map is in the collection of Florissant Fossil Beds National Monument, catalog number FLFO-1013.

Scudder began publishing papers on Florissant as early as 1876, but his culminating work in 1890 was *The Tertiary Insects of North America*, a huge monograph of 734 pages accompanied with beautifully detailed drawings. He later published two other large monographs, in 1893 and 1900, describing the Coleoptera (beetles) from Florissant. The discovery of the extraordinarily rich fossil deposits at Florissant presented Scudder with unique and timely opportunities. In total, he described about 600 species from Florissant. These numbers attest to the overwhelming task that he undertook, and his work was completed with great attention to detail. Scudder worked with both modern and fossil insects and had just completed a catalog of the fossil insects of the world when he began his work at Florissant, and thus he was well poised to undertake the task of examining the fossils from Florissant. He is said to have worked rapidly, and because he sometimes described new species from imperfect specimens, he occasionally made identifications to taxonomic levels that may not have been justified by the characters preserved in the fossils. In 1902, his huge collections were donated to the Museum of Comparative Zoology at Harvard University.

Leo Lesquereux (Figure 9) was to fossil plants what Scudder was to the insects. Both men were pioneers in their work to describe the new fossils coming from the Hayden Survey's discoveries in the American West. Lesquereux came to America from Switzerland in 1847, deaf and with no knowledge of English. Although he never actually visited Florissant, he produced the first scientific publication about the site in 1873, even preceding the publications of Peale and Scudder. In 1878 and 1883, he published lengthy monographs on the Cretaceous and Tertiary fossil plants of the western territories, including descriptions of numerous new species from Florissant. Because of his background, Lesquereux had a strong bias toward comparison with European fossil plants, and for that reason, many of his Florissant identifications were incorrect. Nevertheless, he was the first to name Florissant's fossil plants, and he described more than 100 new species, many of which he illustrated with drawings. The majority of his collections are now at the Smithsonian's National Museum of Natural History.

Figure 9. Leo Lesquereux (1806–1889) was the first to publish on Florissant's paleontology in 1873. Over the course of a decade, he described numerous species of fossil plants. Photograph courtesy of Smithsonian Institution Archives.

Another of the important early scientific collecting parties to Florissant was the Princeton Expedition of 1877. Years later, W. B. Scott (1939, p. 62) recalled his experiences of the journey. "No plan of exploration had been made, no localities suitable for collecting had been fixed; in fact, the expedition threatened to deteriorate into an aimless wandering about. At Florissant, we happened on some fossiliferous beds that afterwards became famous for their beautifully preserved leaves, insects, fishes, and even birds. We made quite extensive collections there and arranged with Mrs. Hill, owner of the land, to forward additional material to Princeton. Our collections of plants were described by Leo Lesquereux, paleobotanist of the Hayden Survey, and the insects by Mr. Scudder, of Cambridge, Mass. We had gathered many species new to science."

Although theirs were among the most significant of the early Florissant collections, the disorganized nature of the Princeton Expedition, along with rivalry between the two leaders, leaves a less than enthusiastic recounting by Scott. "To Osborn, Speir, and myself, the meaningless wandering about Colorado had been one long exasperation, a wanton waste of time, money, and opportunity. . . . Had we remained in Colorado, I do not believe that the experiment of the expedition would ever have been tried again. The collection made at Florissant would have been a very poor return for the great outlay of money and labor." Despite this dismal account, the collections that they amassed during that expedition are *still* among the most impressive and significant fossils to have come from Florissant. In fact, many of these were designated as the type specimens that defined new species, and so it is that the earliest collected are often the most important fossil specimens. Even today, paleontologists must rely on these early collections. The Princeton Expedition collections are no longer housed by Princeton University; in the 1980s they were transferred to the Smithsonian's National Museum of Natural History and the Peabody Museum of Natural History at Yale University.

Scientific studies at Florissant slowed for a decade or two following the early works of Lesquereux and Scudder. This changed with the arrival of Professor T. D. A. Cockerell of the University of Colorado (Figure 10), whose enthusiasm for the area's fossils prompted years of new discovery and inspired research by other specialists. Cockerell organized expeditions to Florissant from 1906 through 1908. Together with other colleagues and his wife, he collected some of the most beautiful fossils ever to come from Florissant. The collections that they amassed were huge, once described by Cockerell as "an almost embarrassing amount of material." His fieldwork was done in cooperation between the University of Colorado, the American Museum of Natural History, Yale University, and the British Museum, and each of these museums still retains portions of his collections. Cockerell was generous with his exchange of specimens, and often the corresponding halves of the same specimen, typically kept together, are instead in different museums. In many instances, half of a specimen is housed at the University of Colorado, whereas the corresponding counterpart is at the American Museum of Natural History or The Natural History Museum, London.

Early in his career, tuberculosis was diagnosed and Cockerell did not expect to live the full life that he did. This is said to explain, at least in part, his incentive for publishing so frequently, and indeed, between 1906 and 1941, he published about 140 papers about Florissant alone. Many of these were quite short, as little as a paragraph in some instances. He studied both plants and animals, often describing new species very briefly and without illustrations. Although the rules for describing new species no longer allow such scanty documentation, it was not unusual in Cockerell's time. Many of the species that he described are photographically documented for the first time in the new Florissant database.

Figure 10. T. D. A. Cockerell (1866–1948) of the University of Colorado organized expeditions to Florissant in 1906–1908, making huge collections that are now in several major museums throughout the United States and Great Britain. He published far more articles on Florissant than any other paleontologist. Photograph courtesy of the T. D. A. Cockerell Collection, Archives, University of Colorado at Boulder Libraries.

Figure 11. Paleobotanist Harry D. MacGinitie (1896–1987) revisited Florissant in 1979, at the age of 83. He made extensive excavations here in 1936 and 1937, and he produced the classic work *Fossil Plants of the Florissant Beds, Colorado,* in 1953. File photograph by Joe Decker, Florissant Fossil Beds National Monument.

A new wave of interest began when paleobotanist Harry D. MacGinitie (Figure 11) came to Florissant during the Great Depression of the 1930s. MacGinitie taught briefly at the University of Colorado, where he met Cockerell. Cockerell encouraged him to revise the old work on Florissant's fossil plants, and to make new collections. Using a horse and plow to remove the overburden of soil and rock (Figure 12), and a hammer and butcher knife to split the shales, MacGinitie excavated three new sites during 1936 and 1937. All of these collections are now at the University of California Museum of Paleontology in Berkeley, where MacGinitie had received his doctoral degree and spent his later years as a research associate.

Following the interruptions of World War II, MacGinitie published the classic monograph *Fossil Plants of the Florissant Beds, Colorado* in 1953. This work was a careful and detailed taxonomic study, and in it, MacGinitie named about 20 new species, but more importantly, he completed the arduous task of revising the older works of Lesquereux and others. At least 40 papers had been written about the Florissant plants since Lesquereux's 1883 publication, and about 260 species had been described. Only about half that number remained valid in his 1953 monograph. This sort of "compression" of names happens when one particular kind of fossil has been given multiple names in earlier literature. During revision, the names are compressed, or synonymized, under only one of these names, usually the first one that appeared in the literature. This name then becomes the only valid name for that fossil species. By this process, MacGinitie completely revised and updated the list of fossil plants, adding some species and deleting others. This is a process that still needs to be completed for the fossil insects, a daunting task, as more than 1,500 species have been described after 130 years of study.

Early paleobotanists such as Lesquereux were concerned primarily with classifying the fossils taxonomically. MacGinitie went further, considering the fossil flora (all of the plants that constituted the ancient vegetation) as a dynamic, evolving plant community and making comparisons with the distribution of modern vegetation. He studied the fossil plants to determine the origin of the plant community, and to make inferences about the ancient climate and elevation. He also considered the effects of taphonomy—those factors that limit and bias the preservation of organisms in the fossil record—on interpretation of the paleoenvironment.

Many other paleontologists—almost too numerous to mention—have left their impact on Florissant. Among the other important historic

Figure 12. MacGinities's excavations at Florissant used horse-drawn equipment to remove the layers of rock. Collecting at Florissant involves considerable effort, and as MacGinitie noted in his 1953 monograph, "more labor is required to obtain a given number of specimens than at any other fossil plant locality the writer has visited." And MacGinitie had visited many localities! Photograph donated to Florissant Fossil Beds National Monument courtesy of Howard Schorn, University of California Museum of Paleontology.

contributors: E. D. Cope (famous for his work on dinosaurs) studied the Florissant fossil fish from 1874 to 1883; W. Kirchner studied the fossil plants in 1898; H. F. Wickham made large collections at Florissant and published numerous studies on fossil beetles from 1908 to 1917; C. T. Brues studied the fossil bees and wasps from 1906 to 1910; F. H. Knowlton published on fossil plants in 1916; M. T. James published on the fossil flies from 1937 to 1941; A. L. Melander published on the fossil flies in 1949; and F. M. Carpenter published on fossil ants in 1930, and in 1992 compiled a two-volume work for the *Treatise on Invertebrate Paleontology* summarizing the fossil insects of the world, including listings for many of the genera found at Florissant. Paul R. Stewart, former president of Waynesburg College, made a large and important collection of Florissant fossils between about 1940 and 1971, assisted in part by that school's summer geology field camp.

More recently, new works have continued to unravel the old secrets of Florissant's past. Steven Manchester has done detailed work showing that at least several of the fossil plants, assigned by MacGinitie to living genera, are actually extinct genera. F. Martin Brown described five new species of insects, two of which are butterflies, a rarity in the fossil record. Emmett Evanoff has revised the terminology for the stratigraphic rock units of the Florissant Formation. Neal O'Brien has used scanning electron microscopy to show details in the fabric of the shale, revealing the processes that preserved the delicate flowers and insects. Various studies by Kathryn Gregory, Jack Wolfe, and me have estimated paleoelevation from the fossil plants, concluding that Florissant was much higher than MacGinitie inferred. In her descriptions of new species of fossil woods, Elisabeth Wheeler has shown that the petrified forest consists of more than just the well-known redwoods. Douglas Nichols and Hugh Wingate, and Estella Leopold and Scott Clay-Poole, have studied the microscopic world of Florissant's fossil pollen. And Dena Smith's examination of insect damage on fossil leaves has provided new evidence about the evolution of plant-insect interactions.

As new work continues, old ideas are constantly reexamined and new interpretations are put forth for future scrutiny. Science is an ongoing process, not just a body of knowledge, and through that process our ideas are continually reevaluated and revised as new discoveries are made and new methodologies are developed. The possible questions still to be asked about the ancient world at Florissant seem endless, and we will probably never have definitive answers for all of them. But surely, future generations of geologists and paleontologists will find plenty to do there.

THE SLOW ROAD TO CONSERVATION AND PROTECTION

By 1915, the Florissant fossil beds area had been mentioned for national monument status, but despite this and subsequent consideration during the following decades, such federal protection was many years in coming. During the 1920s, if not earlier, private landowners were developing the petrified forest for tourists. The area around the Big Stump had been developed by John Coplen as the "Coplen Petrified Forest" by 1922. New stumps were excavated, and visitors were allowed to collect leaf and insect fossils from a site on the property. The old Colorado Midland Railway station was moved to the Coplen site from the town of Florissant, serving as a lodge, which showcased a fireplace made of petrified wood. The site was later operated by the Singer family as the "Colorado Petrified Forest" (Figure 13) from 1927 until the property was acquired by the National Park Service, more than 40 years later. The theft of petrified wood from the property caused the Singers to explore the possibility of a government purchase of their land, and in 1952, they were able to attract a visit from the superintendent of Yellowstone National Park to consider the area as a National Monument.

Another privately owned concession, the "Pike Petrified Forest" (Figure 14), developed adjacent to Singer's Colorado Petrified Forest. This property, located at the site of the Redwood Trio, had been developed by the Henderson family into a tourist attraction by 1922. Later owners of the property, H. D. Miller and John Baker, continued to operate the concession until it closed in 1961. The

Figure 13. The Colorado Petrified Forest was a private concession that featured the Big Stump. It operated from the 1920s through the 1960s. The building that served as the lodge was the old Colorado Midland Railway station, which had been moved to this site from the town of Florissant. This brochure, probably from the 1930s, shows a crowded parking area that attests to the early popularity of this attraction. The site is now a part of Florissant Fossil Beds National Monument, and the old lodge is no longer standing.

building used by Pike Petrified Forest later would become the headquarters for Florissant Fossil Beds National Monument. During the later years of operation, a bitter feud existed between the owners of the Colorado Petrified Forest and the Pike Petrified Forest. Singer and Baker are reported to have used trick signs, tacks in the roadway, and aggressive personal solicitation along the road to detract business from one another. At one point, this competition became so violent that one man was shot in the leg after placing tacks in Singer's roadway.

During the summer of 1956, an unusual visitor stopped at the Pike Petrified Forest. After looking at several of the petrified stumps, he inquired about purchasing one of them, intact, and having it shipped to California for display in his new park. The gentleman then introduced himself as Walt Disney. Following his visit, a large stump was removed with a crane and shipped to Cali-

Figure 14. The Pike Petrified Forest, a rival neighbor and competitor to the Colorado Petrified Forest, was another privately owned enterprise that operated from the 1920s until the early 1960s. After passing through this showy entrance, visitors were able to see the petrified redwood trio and other large trees. The owners apparently were concerned that potential visitors might get sidetracked and visit the Colorado Petrified Forest instead, and they used many tactics to divert attention from their neighbors, including this statement to "be sure the entrance reads Pike Petrified Forest." The site is now the headquarters for Florissant Fossil Beds National Monument.

fornia, where it can still be seen as an attraction in the Frontierland area at Disneyland Park (Figure 15). Walt apparently had expressed some interest in the property during his visit, and the owners of both the Pike Petrified Forest and the Colorado Petrified Forest later offered to sell their properties. A Disney representative responded in 1957 that the company was not interested.

The only private fossil site still operative is the Florissant Fossil Quarry, which has been operated by the Clare family since the early 1950s. They allow visitors to collect plant and insect fossils and have donated many important specimens from

Figure 15. Walt Disney stopped at the Pike Petrified Forest in 1956 and arranged for the purchase of this petrified stump. It was later removed by crane and shipped to California, where it became the centerpiece of this exhibit in the Frontierland area at Disneyland Park. Photograph courtesy of Vincent Santucci. Used by permission from Disney Enterprises, Inc.

their site to museums for scientific research. Many of the most exciting discoveries in recent years have come from this site, and in 1997, Nancy Clare Anderson uncovered the complete, well-preserved remains of a new species of shorebird (Figure 218), one of the more remarkable fossils ever found at Florissant.

Despite several earlier recommendations that Florissant be preserved as part of the public trust, it remained in private ownership through the 1960s. During the late 1960s, real estate developers posed a new threat by putting forth plans to develop a subdivision of A-frame cabins over the fossil beds. Outraged by these plans, concerned scientists such as Estella Leopold and Beatrice Willard, and other citizens, formed the Defenders of Florissant and catalyzed the interest of state and federal legislators. Paleontologists Leopold and MacGinitie provided testimony to the U.S. congressional committee that was considering the issue of establishing the area as a national monument. The local

Figure 16. Following a long legal battle, a portion of the Florissant fossil beds was established as Florissant Fossil Beds National Monument in 1969, ensuring the area's protection from future development. The Monument's logo depicted on the sign is the same specimen of the wasp or hornet illustrated in Figure 206.

land developers and some of the local citizens continued to push aggressively toward the destruction of the fossil beds. Battle lines had been drawn, and the members of the Defenders of Florissant were ready to face the blades of bulldozers, if necessary, to halt the development. Their legal counsel summarized the value of Florissant with the following statement: "The Florissant fossils are to geology, paleontology, paleobotany, palynology, and evolution what the Rosetta Stone was to Egyptology. To sacrifice this 34 million year old record . . . for 30-year mortgages and the basements of the A-frame ghettoes of the seventies is like wrapping fish with the Dead Sea Scrolls." Following a court injunction that temporarily halted the development, both branches of the U.S. Congress passed a bill to establish a national monument from the privately owned lands. On 20 August 1969, President Richard Nixon signed the act into law, providing for the purchase of 6,000 acres and finally creating Florissant Fossil Beds National Monument (Figure 16).

Twenty-five years later, in 1994, the national monument initiated an active paleontology research program when my position as the monument's first paleontologist began. One of the primary efforts has been to inventory all of the published collections of Florissant fossils, which are stored in at least 17 different museums, and to develop a database documenting those specimens along with the publications in which they were described. The outcome of that project has made the idea for this book a reality. Recently, the National Park Service resumed excavations as a part of the resource management program (Figure 17). These new collections include locality and stratigraphic data, which are lacking for most of the specimens in older museum collections. Such information is important for placing the fossils in a three-dimensional context, which in turn will make it possible to answer questions relating to changes in the paleoecosystem through time and space.

This chapter has shown how the Florissant fossils first came to the public's eye, introduced the paleontologists who made the first fossil discoveries here, and summarized what has been done to protect and conserve these paleontological treasures. But this history—a mere 130 years—is dwarfed by the depth of time that we must fathom in order to return to Florissant's ancient world of 34 million years ago. We turn now to examine this very different time at Florissant, one in which volcanoes erupted, a bygone community of plants and insects thrived, and the entire world was much warmer.

Figure 17. Modern scientific excavations at Florissant involve digging trenches to expose the rock layers, and each layer is carefully examined and measured. This type of information is important for showing how the fossil composition changed through time. The plastic tarp is used to protect the easily weathered shales from the heavy afternoon thunderstorms that occur almost daily during the summer.

RELATED REFERENCES
History of early fieldwork: Kohl and McIntosh (1997); Scott (1939)
Scudder biography: Kingsley (1911); Mayor (1924); Morse (1911)
Lesquereux biography: Andrews (1980)
Cockerell biography and bibliography: Weber (1965, 2000)
MacGinitie memorial biography: Wolfe (1987)

GEOLOGIC SETTING AND PROCESSES OF FOSSILIZATION

The rocks of central Colorado preserve the story of geologic events spanning 1.7 billion years of Earth's history. Against this expanse of time, Florissant's age of 34 million years seems young. Ninety-eight percent of the region's recorded geologic history had already elapsed by the time Florissant's fossils were laid to rest (Figure 18).

Proterozoic rocks of the region consist of granites that formed from cooling magma deep beneath the earth's surface, and of metamorphic rocks that formed as older rocks were altered by intense heat and pressure. These rocks form the basement upon which rest the sedimentary rocks of the Paleozoic, Mesozoic, and Cenozoic Eras.

During the first part of the Paleozoic Era, in the Cambrian through Mississippian Periods, limestones, shales, and sandstones were deposited in shallow seas that covered central Colorado. These rocks often contain marine invertebrate fossils, and occasionally fish and plant remains. During the later part of the Paleozoic, in the Pennsylvanian and Permian Periods, mountain building and uplift formed the "Ancestral Rocky Mountains." The older rocks contained within these uplifts were then weathered and eroded into sediments that were redeposited in surrounding basins along the flanks of the uplifts. Layers of these colorful sedimentary rocks are boldly exposed today at places like Garden of the Gods, along the face of the Front Range near Colorado Springs.

During the early part of the Mesozoic Era, in the Triassic Period, the Ancestral Rockies were eroded to form an upland with low relief. The Morrison Formation, famous for its numerous dinosaur fossils, was deposited by streams during the Jurassic Period. By the end of the Jurassic, the Morrison Formation had completely buried

Left: Detail of a tsetse fly. Courtesy of the American Museum of Natural History.

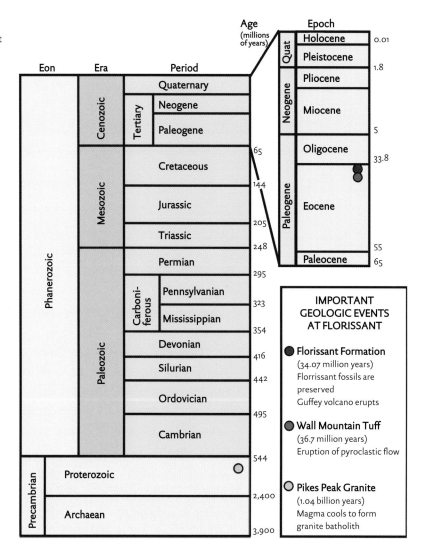

Figure 18. Geologic time chart showing important events in the Florissant area. Boundary ages are based on Haq and Van Eysinga (1998).

the topography of the remnant Ancestral Rockies. As the region subsided during the Cretaceous Period, a shallow interior seaway invaded the central part of the continent. Numerous formations of sandstone and shale were deposited in this interior seaway until late in the Cretaceous, burying the region under an additional 3,000 meters of sediment. Near the close of the Cretaceous, about 65–70 million years ago, a new mountain-building episode known as the Laramide Orogeny began. For several million years, this orogeny compressed the earth's crust within the continental interior, buckling the older rocks and uplifting them along faults to begin forming the modern Rocky Mountains.

The Laramide Orogeny continued into the early part of the Cenozoic Era. By the late Eocene Epoch, volcanism began in the region and continued spo-

Figure 19. The oldest rock exposed in the immediate vicinity of Florissant is the 1.04-billion-year-old Pikes Peak Granite. The granite formed when molten magma slowly cooled and crystallized deep beneath the earth's surface. It was uplifted several hundred million years later during various phases of mountain-building activity. During the late Eocene Epoch 34 million years ago, when Florissant's fossil plants and insects lived, the Pikes Peak Granite formed the hilly landscape around the ancient Florissant valley much as it does today.

radically throughout the Oligocene Epoch. It was this volcanism that set the stage for the fossil preservation of the ancient Florissant ecosystem. Florissant's time had finally come.

MIGHTY VOLCANOES, GIANT TREES, AND TINY INSECTS: MAKING THE FOSSILS OF FLORISSANT

The oldest rocks around Florissant are Proterozoic granites. Pikes Peak Granite, 1.04 billion years old, makes up nearby Pikes Peak, towering 4,301 meters, and is also exposed in the slopes and low hills around the Florissant valley (Figure 19). The granite formed from magma that cooled deep beneath the earth's surface during the Proterozoic, forming a large intrusive body known as a batholith. It was uplifted during the mountain building of the Laramide Orogeny, beginning in the late Cretaceous about 65–70 million years ago. Florissant is located along the axis of a north-south uplift, flanked to both the east and the west by large basins. As the uplift progressed, older Paleozoic and Mesozoic rocks were eroded from the uplifted area and redeposited as sediments into the surrounding Denver and South Park basins. The ancient granite, which formed the core of the uplift, had been exposed at the surface. By the late Eocene, 37 to 34 million years ago, continued erosion had beveled the landscape to form a surface of low to moderate relief, and stream drainages had been cut into the granite. The evidence from fossil plants suggests that by this time, the Florissant area may already have been uplifted to its present elevation.

The scene began to change during the late Eocene as volcanoes unleashed

Figure 20. Remnants of the Wall Mountain Tuff can be seen in rock outcrops around the modern Florissant valley. The Wall Mountain Tuff formed 36.7 million years ago when a superheated pyroclastic flow—a "volcanic hurricane"—ravaged the ancient Florissant valley. The eruption originated from a volcanic caldera 80 kilometers west of Florissant.

fiery havoc across the landscape. First came a cataclysmic eruption from near present-day Mount Princeton, about 80 kilometers west of Florissant. This event was the culmination of a long process, which began when molten magma was emplaced beneath the surface and caused fractures to develop in the earth's crust. The fractures eventually tapped the magma chamber, releasing an enormously powerful and voluminous eruption known as a pyroclastic flow. At the source of this eruption, an immense craterlike depression known as a caldera formed when the eruptive center collapsed into the depleted magma chamber following the eruption. Although this caldera was large, it is no longer evident in the modern topography. The pyroclastic flow consisted of a superheated cloud of ash, pumice, mineral crystals, and glass, all suspended in hot gases and behaving as a flow. Moving at speeds of 160 kilometers per hour or more, the pyroclastic flow raced across the landscape to the east, generally following drainages as it swept through the ancient Florissant valley like a volcanic hurricane. Almost instantly, the Florissant area was devastated by one of nature's most violent acts. Plant and animal life in the path of the eruption was destroyed. As the pyroclastic flow came to rest, it compacted and the hot particles quickly fused together and cooled, forming a rock known as welded tuff. This particular rock unit, the Wall Mountain Tuff, is dated at 36.7 million years and is found in scattered outcrops throughout the Florissant valley (Figure 20), and as far away as

Figure 21. This reconstruction illustrates the impact of the ancient Guffey volcano on the Florissant valley. Lahars originated when volcanic debris on the slopes of the volcano became saturated with water and flowed into the valleys below. The muddy outflow from one of these lahars, rich in ash and pumice, buried the fossil forest at Florissant. Another lahar, laden with large boulders, flowed down an ancient side tributary and came to rest near the confluence with the main Florissant valley, where it formed a natural dam (left center) and created Lake Florissant. Occasionally, explosive volcanic eruptions covered the landscape with ash, some of which washed into the lake and was deposited as sediment. Original artwork produced by Linda Lutz-Ryan, with geological consultation from Emmett Evanoff.

Castle Rock (between Colorado Springs and Denver), about 150 kilometers from its source. It was a bad day for late Eocene Colorado.

About 2 million years later, another type of volcano developed much closer to Florissant. It towered above the landscape as a cluster of large, coalesced stratovolcanoes, similar to peaks such as Mount Rainier, Mount Saint Helens, and Mount Shasta in the modern Cascade Mountains. Known as the Guffey volcanic center, this ancient volcano was situated within a larger area known as the Thirtynine Mile volcanic field, and was located about 25–30 kilometers southwest of the ancient Florissant basin. Eruptions from the Guffey volcano were similar to those that brought fame to Mount Saint Helens in 1980. These eruptions included domes, lava flows, and pyroclastic breccias that formed the volcano itself; explosive pyroclastic eruptions that showered ash and pumice across the landscape; and lahars (volcanic debris flows) that came off the slopes of the volcano as broad sheets and flowed down the ancient valleys (Figure 21). The ash and pumice consolidated into a type of rock known as tuff, and the lahars formed rocks ranging from mudstones to boulder-size conglomerates that

Figure 22. The deposit shown here formed when a lahar flowed down a tributary and into the ancient Florissant stream valley. When it came to rest, the lahar created a natural dam, which formed Lake Florissant. The remains of this lahar can be seen in a roadcut several kilometers south of Florissant. The rock consists of a muddy matrix containing cobbles and boulders of volcanic rocks from the Guffey volcano, along with much older pieces of granite that were picked up along the valley bottom as the lahar flowed. This sort of mixture, with rocks of different kinds and sizes, is typical of lahars.

Figure 23. The hills along the skyline in this photograph show what little remains today of the Guffey volcano and the surrounding Thirtynine Mile volcanic field. The low hills in the saddle along the horizon (to the left of center, above the stream) consist of the core of this deeply eroded stratovolcano, and the higher buttes are remnants of lahars and lavas that flowed off the flanks of the volcano. This view along modern Fourmile Creek is approximately the route of one of the ancient drainages that some of the lahars followed, including one lahar that flowed past this spot and into the confluence with the ancient Florissant stream valley (several kilometers behind the viewpoint here), creating a dam and causing Lake Florissant to form. The view looks west from a point along the road between Evergreen Station and Guffey (see Figure 24).

were composed of volcanic debris and other sediment (Figure 22). Repeated eruptions from the Guffey volcano had a long-lasting and major impact on the landforms and deposition in the ancient Florissant valley. Today, the Guffey volcano has withered from its late Eocene majesty and is little more than remnant hills exposing the guts of a long-dead and much-eroded volcano (Figure 23). But during its reign over the ancient landscape, the fossils of Florissant were preserved.

How can something as delicate as the body of an insect survive something as destructive as a volcano and become preserved in the fossil record? The processes of fossilization at Florissant were complex, but ironically, without the volcano, the remains of the insect would have deteriorated. It is not unusual that certain types of volcanic events—including those that shower the earth with air-borne volcanic ash or inundate valleys with lahars—can leave their impact on the fossil record. And so it was with the ancient Guffey volcano. All the fossils at Florissant, from the largest redwood to the smallest insect, would never have been preserved without the influence of the Guffey volcano.

Between 34 and 35 million years ago, a forest of large redwood trees grew in volcanic soils along a stream that flowed through the Florissant valley. On the higher slopes along this ancient valley, vegetation adapted to drier conditions grew among rock outcrops of eroded Pikes Peak Granite and Wall Mountain Tuff. The main stream flowed to the south-southeast and was joined by several tributaries along its course. One of these, in the southern part of the valley, extended westward into headwaters on the slopes of the Guffey volcano. Loosely consolidated volcanic ash and larger debris covered the barren flanks of the volcano following eruptions. This material became saturated by rainfalls and possibly melting snow, either during or shortly following an eruption, and formed lahars that flowed like massive slurries of wet concrete down the slopes of the volcano and into valleys (Figure 24). They traveled at speeds of 5 to 140 kilometers per hour, moving fastest along the steep flanks and more slowly in lowland areas. They often picked up large boulders and other debris along the way and covered the valley bottoms in blankets of volcanic sediment about two to six meters deep. As the boulder-laden flow slowed, an outflow of water-rich muddy slurries continued to carry finer sediments beyond the front of the flow.

One of these lahars flowed into the ancient Florissant valley and entombed the lower portions of the giant tree trunks in about five meters of volcanic mud and ash. The lahar hardened and consolidated to form a volcanic mudstone. Over time, the bases of these trees—those parts that were buried in the lahar—were petrified by permineralization. This process happened as groundwater, enriched in dissolved minerals from the volcanic rocks, permeated the wood and

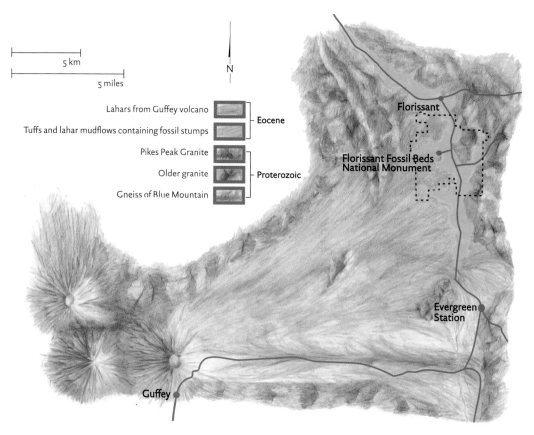

Figure 24. An artistic reconstruction depicts an aerial view of what the ancient Florissant valley might have looked like near the end of the Eocene Epoch just over 34 million years ago. The Guffey volcanic center consisted of perhaps three closely coalesced stratovolcanoes (lower left) that sent torrents of wet lahars down their flanks and into the valleys below. These lahars carried large boulders of volcanic and other debris, and as they slowed, muddy slurries continued to carry finer sediments into the Florissant valley, burying and fossilizing the bases of giant redwood trees. The road that now follows from Evergreen Station to Guffey follows up an ancient tributary of the main Florissant valley. This eastward-flowing tributary was filled by the lahars, and near the confluence with the southward-flowing Florissant valley, the lahars piled up to form a natural dam. The waters of the main Florissant valley filled behind this dam to create Lake Florissant, and it was in this lake that the leaf and insect fossils were preserved. Original artwork produced by Linda Lutz-Ryan, with geological consultation from Emmett Evanoff.

precipitated silica as a mineral within the cell walls. The lower parts of these huge trees—now the petrified stumps that contribute to Florissant's fame—were literally turned to stone. The lahar had preserved the stumps of the giant redwoods, but not the foliage or tiny insects. But conditions were about to change, creating new situations in which leaves and insects could enter the fossil record as well.

More lahars flowed down the valley of the southern tributary, from headwaters near the flanks of the Guffey volcano (Figure 24). These lahars carried sediments ranging in size from mud to boulders, mostly of volcanic origin but also including boulders of the older granite that lay along the valley floor (Figure 22). One of these lahars, or a series of lahars, formed a natural dam at the confluence of the southern tributary with the main Florissant drainage. This dam impounded the southward-flowing Florissant drainage and created a lake about 1.5 kilometers wide and 20 kilometers long to the north (Figures 21 and 24). The shape of ancient Lake Florissant resembled a modern man-made reservoir with drowned tributaries forming distinct arms around the margin of the lake. The modern outcrop area of the Florissant Formation (Figure 25) preserves this pattern remarkably well, indicating that there has been minimal structural change in this area since the Eocene.

Like modern stratovolcanoes, the Guffey volcano was active sporadically over a long period, perhaps several hundred thousand years. Times of quiescence were punctuated by outbursts of activity. Decades or even centuries may have elapsed between eruptions. By analogy, Mount Saint Helens is known to have erupted about 10 times during the 1800s, but its history shows that there were also periods of inactivity lasting several thousand years. Throughout its long life, the Guffey volcanic center produced many lahars and andesitic lava flows, which eventually buried much of the surrounding landscape near the volcano. Explosive eruptions of ash and pumice affected areas even more distant from the volcano itself. Other volcanoes in the region, such as the Mount Aetna and Grizzly Peak calderas to the west, also may have been active during this time and contributed to the ash and pumice falls in the Florissant valley. Some of the eruptions were major, depositing layers of ash a centimeter or more in thickness, but others were minor, just dusting the surface. Some of this ash was deposited directly into Lake Florissant, but much of it fell across the landscape and later washed into the lake. Volcanic ash has the same silica-rich composition as glass, and lacking a crystalline structure, it is unstable, allowing the silica to dissolve gradually as it weathers. Scanning electron microscopy reveals that some of this ash began to weather into clay before or during its deposition in the lake. As this weathering occurred, the chemistry of the lake water became enriched with the dissolved silica, and a thin layer of ash-clay was deposited on the lake bottom.

Diatoms, which are microscopic algae with hard siliceous shells known as frustules, lived within the waters of the lake. The abundance of silica allowed the diatoms to periodically "bloom" into abundant populations. Diatom blooms were followed by massive die-offs, caused during periods when overpopulation depleted the lake waters of silica or other nutrients. As the diatoms died, bil-

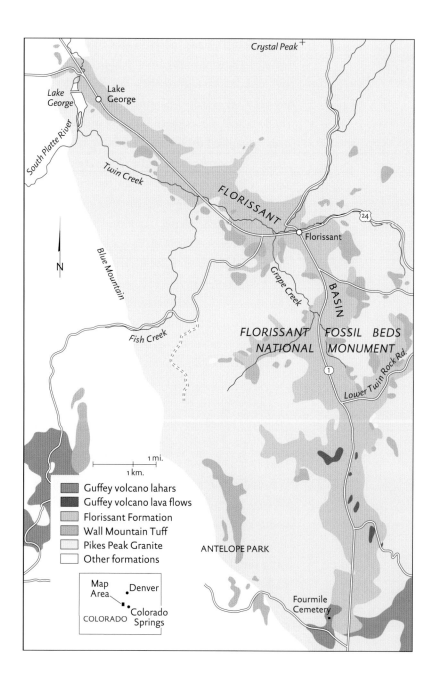

lions upon billions of their microscopic siliceous frustules settled to the lake bottom, forming a thin layer above the layer of ash-clay.

Following this, another layer of the weathered ash-clay was deposited, and the cycle repeated itself. Each sequence of ash-clay and diatoms is referred to as a couplet. The repetition of these couplets was caused by cyclic events in the lake. Such events were initiated when runoff from rainfall washed some of the

Figure 25. Geologic map showing the distribution of the Florissant Formation and surrounding older rocks. The narrow, light-brown area extending the length of the map represents the Florissant Formation, consisting of 34.1-million-year-old lake-deposited shales and tuffs (volcanic ash and pumice), stream-deposited sediments, and volcanic lahars. Notice that the irregular shape of the Florissant Formation preserves inlets of the ancient lake, which formed from drowned tributaries when the stream valley was flooded to form the lake. The medium-brown areas in the southern part of the map show rocks that formed from the eruptions of the Guffey volcano (not visible on the map), including the lahars that dammed the ancient drainage to form Lake Florissant. The dispersed patches of pinkish mauve depict the remnants of Wall Mountain Tuff, which is a 36.7-million-year-old welded tuff that formed when a caldera about 80 kilometers west of Florissant erupted. The area covering much of the central and eastern part of the map shows the local distribution of Pikes Peak Granite, a 1.04-billion-year-old intrusion of granite, which is much more widespread to the east of this map. Most of the cream-colored areas in the western half of the map consist of Precambrian rocks older than the Pikes Peak Granite. This map is based on a portion of U.S. Geological Survey Map I-1044 (Wobus and Epis 1978).

weathering ash into the lake, where it continued to weather into clay and increased the silica concentration within the lake. This, in turn, helped stimulate diatom blooms. It is unclear, however, whether these couplets represent truly annual cycles, or cycles of another periodicity that may have been influenced by multiple rainfall events within a year. It is also possible that diatom blooms occurred more or less often than once a year.

The microlayers of ash-clay and diatom couplets compacted to form "paper shale," so named because it splits to form paper-thin sheets. It is these paper shales that contain the extraordinarily well preserved plant and insect fossils, which are revealed as the thin layers of shale are split. The rock can be classified as shale because of its clay content and laminated character, but most precisely it can be called a diatomaceous tuffaceous shale to denote the significant components of diatoms and ash. The layers of paper shale are composed of numerous couplets of the alternating diatom and ash-clay laminae, each about 0.1 to 1.0 mm in thickness. The couplet-forming events were repeated many times before the pattern was interrupted (Figure 26).

Occasionally, the couplet-forming paper shale cycle of sedimentation was interrupted, briefly punctuated either by outbursts of eruptions that produced more voluminous amounts of coarse ash and pumice, or by a period in which increased runoff brought more volcanic sediment into the lake. This ash and pumice was deposited in the lake as thicker layers of tuff, alternating at irregular intervals with the paper shale layers. There were thus two different patterns or cycles of sedimentation that caused the lake deposits to accumulate: diatoms and ash-clay, and volcanic ash eruptions. These different cycles of sedimentation can be observed at various scales, from an outcrop of the lake deposits in a hillside to the minute details seen in scanning electron microscopy (Figures 27–31).

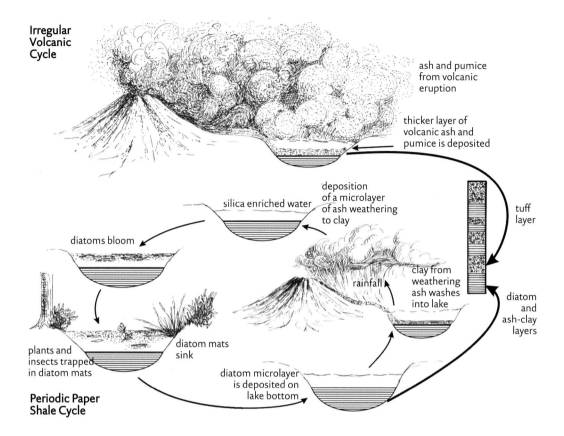

Figure 26. At least two different cycles of sedimentation caused the formation of the Florissant lake deposits. The lower cycle in this diagram shows the formation of the fossiliferous paper shale, which formed when the microscopic frustules of freshwater diatoms settled to the bottom of the lake as mucilaginous mats. These microlayers of diatoms alternated with microlayers of clay formed by the weathering of volcanic ash as it washed into the lake from the surrounding landscape. Together, these two microlayers formed a couplet (shown here as a band of white diatoms and a band of dark ash-clay) about 0.1 to 1.0 mm in thickness. This cycle was periodical and repeated many times before it was interrupted by the other component of sedimentation, illustrated here in the upper cycle, in which sporadic volcanic eruptions produced massive quantities of ash and pumice. These layers were much thicker and formed more rapidly than the paper shale. The stratigraphic column to the right shows how these complex cycles interacted to form the layers of rock at the bottom of the lake. This illustration is diagrammatic only, and the thickness of the layers as illustrated is not to scale. Drafted by Linda Lutz-Ryan.

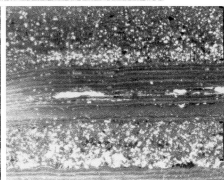

Figure 27. This cut in a hillside reveals the layers that formed when volcanic sediments from the Guffey volcano, along with clay and diatoms, accumulated at the bottom of Lake Florissant and compacted to form flat-lying beds of rock. These beds vary in thickness, from the thin laminations of the fossiliferous paper shales to the thicker layers of volcanic tuff (ash and pumice) and siltstone. The thin shale layers formed slowly over many decades, whereas the layers of volcanic ash accumulated much more rapidly. This photograph shows the privately owned site at the Florissant Fossil Quarry in the town of Florissant.

Figure 28. Close examination of the Florissant shale in cross-section reveals alternating layers of volcanic ash and paper shale. The paper shale, so named because its layers are nearly as thin as paper, is evident as the darker areas of clay containing extremely thin bands of white diatom layers (not discernible at the scale of this photograph). The thicker light gray areas formed from volcanic ash that was deposited during more active pulses of volcanism. File photograph, Florissant Fossil Beds National Monument.

Figure 29. Magnification of a thin section through the shale shows two alternating cycles of sedimentation. The coarse layers represent tuff (volcanic ash) that was deposited rapidly during eruptions of the Guffey volcano. The layers of ash in this illustration are quite thin, although other layers are as thick as several centimeters. Between the two tuff layers is another layer showing the banded cycles of paper shale deposition, which occurred very slowly. The paper shale cycle consists of light bands formed by diatom deposition, alternating with dark bands of clay formed when weathered volcanic ash washed into the lake. Photograph by Neal O'Brien, State University of New York at Potsdam. Reproduced with permission of *Rocky Mountain Geology,* University of Wyoming, Laramie. Scale: × 8

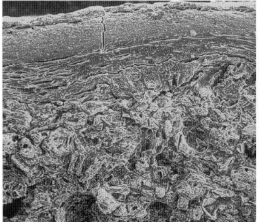 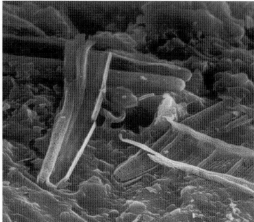

Figure 30. Scanning electron microscopy takes us yet one step closer in magnification than the previous three figures. This cross-section shows a layer of volcanic ash (coarse layer in the bottom half), a thin layer of diatoms in the middle (appearing as parallel lines in profile), and a layer of ash weathered to clay (fine layer at the top). Photograph measures 2.2 × 1.8 μm (micrometers). Photograph courtesy of Neal O'Brien.

Figure 31. By the billions, the tiny frustules of diatoms accumulated on the floor of Lake Florissant during a diatom die-off. Many of the diatoms may have sunk to the bottom of the lake en masse after forming mats near the surface. Photograph measures 20.8 × 17.5 μm. Photograph courtesy of Neal O'Brien.

Whereas it took decades to form a few centimeters of the shale, a similar thickness of the tuff could have been deposited in only a few hours or days. Although fragments of wood and other large fossils occasionally are found in the tuff layers, these conditions were not suited for preserving the kinds of delicate organisms that are found in the paper shale layers.

Much of the early and popularized literature on Florissant attributes the preservation of leaves and insects to catastrophic eruptions of ash from the Guffey volcano. These sensationalized accounts describe boiling lake water, organisms suffocating in poisonous fumes, insects dropping in massive clouds of ash to become trapped in the lake, and butterflies struggling in vain to avoid their inevitable fossilization. Although it is true that cataclysmic eruptions sporadically showered the lake with large amounts of ash and pumice, and that a lahar buried a forest, these were not events conducive to the preservation of delicate insect or flower fossils. In fact, most of the ash and pumice layers are sparsely fossiliferous or entirely devoid of fossils because they were deposited too rapidly and are too coarse in texture to incorporate and preserve the small parts of plants and insects.

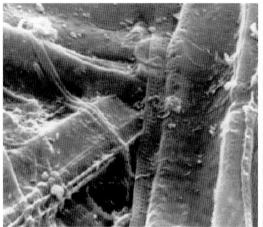

Figure 32. Scanning electron microscopy reveals remarkably well preserved mucilaginous strands exuded from the pores of fossil diatoms. The diatoms secreted this mucus as they became stressed, and as billions of diatoms did so, they began to cling together to form large mats near the surface of the lake. Photograph measures approximately 12.6 × 11.0 μm. Photograph courtesy of Neal O'Brien.

Figure 33. Scanning electron microscopy shows diatoms in a mucilaginous film on the surface of a fossil. The arrow indicates the thin film of mucus overlapping the frustule of a fossil diatom below. This film of mucus was probably part of a much larger diatom mat that entrapped the fossil leaves and insects, carrying them to the bottom of the lake and aiding in their preservation. Photograph measures approximately 48 × 41 μm. Photograph courtesy of Neal O'Brien.

Analysis of the paper shale under the high-powered scanning electron microscope has disproved the cataclysmic scenario and has shown instead that most of the small plant and insect fossils were preserved under conditions of slow deposition during relatively quiet intervals between volcanic activity. It now appears that microbial processes associated with diatom and bacterial activity were important contributors to the preservation of fine details during fossilization. Studies in which modern diatoms are grown in the laboratory show that when diatoms become stressed (for example, from too much competition for nutrients), they secrete a polysaccharide mucilaginous "slime" that causes them to aggregate, forming thin mucus mats that float near the surface of the water. Associated bacterial colonies also may contribute to the production of mucus in these mats. Evidence of similar mats can be seen in the Florissant shale under high magnification (Figures 32 and 33), and close examination of the shale has given rise to the hypothesis that densely packed diatoms formed an entangling mucus mat that enveloped dead leaves and insects at or near the surface of Lake Floris-

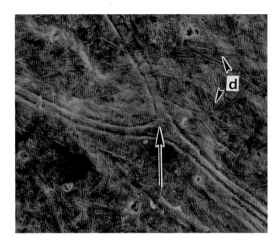

Figure 34. Diatoms (indicated by "d") cover the surface of this insect wing, which was probably enveloped in a diatom mat when it sank to the bottom of Lake Florissant. The diatom mat served both as a mechanism for transporting the insect to the bottom and as a protective coating that retarded the decay of the insect's body. Recent research suggests that these diatom mats were largely responsible for the exceptional preservation of the Florissant fossils. The conspicuous forking lines indicated by the large arrow are veins in the insect's wing. Photograph measures approximately 30 × 29 μm. Photograph by Neal O'Brien. Reproduced with permission of *Rocky Mountain Geology,* University of Wyoming, Laramie.

sant. Scanning electron microscopy confirms that the diatom frustules are closely intertwined with the appendages and wings of insect fossils (Figure 34). The protective coating provided by the mats is thought to have prevented rapid biological decay and chemical decomposition of the enclosed organisms, forming a wrap similar to the way in which plastic bags are used to protect food. As the mats increased in size, they became heavy and rapidly settled, carrying the entrapped organisms to the lake bottom. Here, they came to rest on top of the already accumulated thin layer of ash-clay, and formed the thin diatom layer of a paper shale couplet. This, in turn, was soon buried beneath another thin layer of ash-clay, and the process repeated itself.

Incorporated into these sediments, the leaves and insects were flattened as they collapsed from decay and were pressed by the weight of the overlying water and new sediment accumulating on top of them. Once buried, some of the organisms did decay, leaving only an imprint of their form. These became impression fossils as the sediment compacted and hardened to form shale. Other

GEOLOGIC SETTING AND PROCESSES OF FOSSILIZATION   37

Figure 35. This beetle clearly shows the differences that can be seen when the two corresponding halves of a specimen are compared. The illustration on the left shows the insect's dorsal side, and the one on the right is the ventral side. Specimen NHM-IN.26241, courtesy of The Natural History Museum, London.

Scale: × 4.4

organisms decayed to leave a residue of organic carbon in the sediment. These became compression (or carbonization) fossils, typically darkened in color by the organism's residual carbon. In many of the fossils, especially insects, the wings are preserved as an impression, and the rest of the body as a compression. Splitting the thin layers of shale exposes the fossils as two corresponding halves. It is often possible to see characters of the insect's dorsal (back) side on one slab, and those of the ventral (bottom) side on the other slab (Figure 35). Preservation in the diatom layers involves a type of compaction that still retains some of the three-dimensional characteristics of organisms, including some insect organs and appendages as well as structures *within* leaves. This evidence indicates that the diatom mats played a crucial role in the exceptional preservation of the Florissant fossils.

And so it was that the Guffey volcano triggered the processes and conditions that preserved the fossils. It buried the bases of giant trees in volcanic mud. It created a dam, which formed a lake. It dusted the landscape with volcanic ash, which washed into the lake and nourished the diatoms with silica. The diatoms, in turn, were key agents in the exquisite preservation of delicate insects and leaves. In spite of its great power, the Guffey volcano had facilitated the preservation of delicate insects and flowers gently and slowly, not violently.

LAYERS OF ASH AND SHALE: THE PAGES OF TIME

During its long period of activity, the Guffey volcano left its mark in the ancestral Florissant valley. The history of Florissant's ancient world—both the life and the volcanoes—was recorded layer by layer on the bottom of Lake Florissant. Eventually, volcanic deposits—lahars, ash, and pumice—completely filled the basin of the ancient lake, consolidating along with the microscopic layers of diatoms to form distinguishable layers of rock that record Florissant's geologic history. These deposits are grouped into units that can be mapped and described, and from this the nature and sequence of geologic events can be deciphered.

The stratigraphic column (Figure 36) shows the four major rock units at Florissant: the Pikes Peak Granite, of Proterozoic age; and the Wall Mountain Tuff, Tallahassee Creek Conglomerate, and Florissant Formation, all of late Eocene age. The Florissant Formation has been further subdivided by Evanoff into six informal units: the lower shale unit, lower mudstone unit, middle shale unit, caprock conglomerate unit, upper shale unit, and upper pumice conglomerate unit.

The shale units of the Florissant Formation represent lake deposits, and three such units can be distinguished. Each of these shale units consists of sequences of paper shale (formed by the couplets of diatoms and ash-clay) alternating with tuffs (formed from ash and pumice) (Figure 28). The vast majority of Florissant's fossils come from these paper shale layers. The lower mudstone unit and the caprock conglomerate unit form the boundaries between the three shale units (Figure 36). The lower mudstone consists of stream deposits and, at the top, the consolidated deposit of a muddy lahar that buried the trees to form the petrified forest. The lower mudstone unit was deposited along the valley bottom, not in a lake. This raises the intriguing question of why there are two lake-deposited units separated by stream and mudflow deposits. One explanation is that there may have been two generations of Lake Florissant. In this scenario, the earlier lake is represented by the lower shales, which are very fossiliferous and presently outcrop in the immediate vicinity of the town of Florissant. This lake formed when a stream valley flooded behind a lahar dam several kilometers to the south of Florissant. Layer by layer, sediments began filling the lake. Eventually, the lake either was completely filled by these sediments or was drained as the lahar dam eroded away. The filled surface of the old lake became the floodplain of a new stream valley. On this floodplain, weathered ash and pumice from the surrounding landscape were deposited by the stream to form part of the lower mudstone unit. Later, a lahar flowed into the valley, partially burying the redwood forest and forming the upper part of the lower mudstone

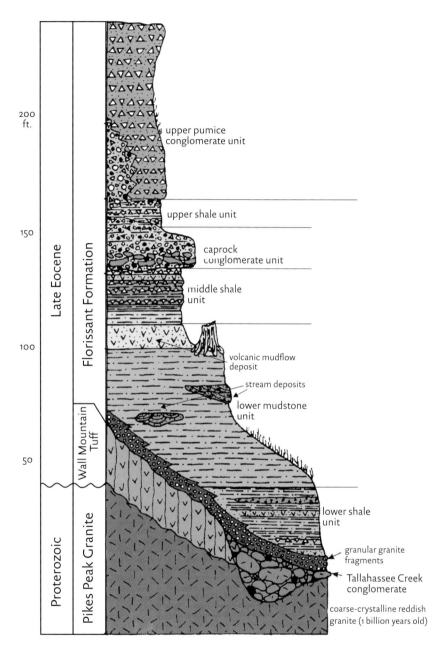

Figure 36. The rocks exposed in the Florissant area are illustrated as a statigraphic column. The oldest rock is the Pikes Peak Granite, of Precambrian (Proterozoic) age more than one billion years old. Younger rocks of the Paleozoic and Mesozoic Eras had already eroded from the Florissant area by the late Eocene Epoch, when the Wall Mountain Tuff was emplaced as the result of a pyroclastic flow erupted from a distant caldera. The Florissant Formation contains the famous plant and insect fossils and is divided into six units. These units were deposited in a lake and in a stream valley, and their composition includes shales, mudstones, conglomerates, and volcanic tuffs. Many of the sediments in the Florissant Formation originated from the Guffey volcano. The stratigraphy is based on Evanoff et al. (2001). Diagram provided courtesy of Emmett Evanoff.

Figure 37. The caprock conglomerate formed when a lahar or debris flowed into the waters of Lake Florissant. As this lahar settled and compacted on the lake bottom, water was squeezed out of the sediment as if from a sponge, leaving water-escape tubes, which are preserved as vertical lines in the caprock conglomerate.

unit. Another lahar, possibly contemporaneous with the one that buried the forest, again formed a natural dam several kilometers to the south, initiating a second, younger generation of Lake Florissant. Both the middle and upper shale units were deposited in this second lake. The middle shale, well exposed within Florissant Fossil Beds National Monument, is very fossiliferous and includes many of the historic sites where Scudder, Cockerell, and MacGinitie collected.

The caprock conglomerate unit, separating the middle and upper shale units, originated when a lahar or debris flow actually entered the waters of Lake Florissant and was deposited on the lake bottom. Evidence of this origin comes from water-escape tubes in the caprock conglomerate (Figure 37). These formed as the water-saturated debris flow came to rest and was compacted on the lake bottom, and water was squeezed out of the sediment. The caprock conglomerate is so named because it forms a hard layer that caps the middle shale unit, protecting it from erosion. Above the caprock, further deposition in the lake formed the upper shale unit. Following an interval of increased volcanic activity, the upper pumice conglomerate unit formed as streams flowing from the west washed large amounts of pebble-size pumice into the lake basin. The lower portion of this unit was deposited in the lake, as indicated by the presence of clam shells, whereas cross-bedded structures in the upper portion are evidence of stream deposition. This series of eruptions may have been the event that ended the existence of Lake Florissant.

The only attempt to estimate the duration of Lake Florissant's existence has been by McLeroy and Anderson. They estimated the total number of couplets in the shale, and assumed that each couplet represents an annual cycle. This method gives a crude estimate suggesting that the lake lasted for about 2,500 to 5,000 years. The only caveat to this estimate is that each couplet may not, in fact, represent a cycle that is annual.

Most of the rocks that later covered the Florissant Formation before the beginning of the Pleistocene Epoch have since been stripped by erosion. Some remnants do persist, however, including later volcanic deposits of andesitic lava flows and breccias from the Guffey volcanic center. During the later part of the Tertiary Period, sand and gravel deposits formed a layer that is now exposed as remnants several kilometers to the east of Florissant, near Divide.

During the Pleistocene ice age, glaciers did not extend into the Florissant valley but were present only on the higher reaches of Pikes Peak. At Florissant, Pleistocene deposits consist of sand and gravel composed of weathered Pikes Peak Granite, volcanic particles, and mud. These deposits, which formed from debris accumulation along the bases of slopes and from sediments washed into the valley by streams, form a thin layer above much of the Florissant Formation. Mammoth bones have been found in these sediments, including a molar with an associated portion of the jaw. This bone has been radiocarbon dated at 49,830 ± 3,290 years B.P. (considered as greater than 43,250 at two standard deviations). This is the upper limit of reliability for radiocarbon dating and therefore represents a minimum age for the mammoth.

THE BOUNDARIES OF TIME: HOW OLD IS FLORISSANT?

The casual reader of earlier literature on Florissant could easily become confused about the age of these fossils, and for good reason. Throughout the long history of study at Florissant, various workers have considered the age of the Florissant deposits to be either Pliocene, Miocene, Oligocene, or Eocene. MacGinitie made the first reliable age determination in 1953, concluding that the age of the fossils was early Oligocene. He reached this conclusion by comparing the Florissant fossil plants with those from other fossil floras, and by considering the fragmentary evidence of fossil mammals, including an extinct horse and oreodont. More recently, fragments of a brontothere have been found. These fossil mammals are known to have lived together only during a restricted period in earth history, which is known as the Chadronian NALMA (i.e., North American Land Mammal Age, based on a separate timescale defined exclusively by fossil mammals). In MacGinitie's time, the Chadronian was considered to belong in the Oligocene Epoch. For a long time, geologists considered the boundary between the Eocene and Oligocene epochs to be at 38 million years. During the 1980s, however, researchers used new evidence from Italy and elsewhere to define this boundary more precisely. As a result, the boundary is now placed at 33.7 to 33.9 million years ago, and the Chadronian NALMA is now shown to belong in the late Eocene. Radiometric dates $^{40}Ar/^{39}Ar$ from the Florissant Formation give an age of 34.07 million years. Our concept of Florissant's age didn't really change, but our concept of where, precisely, to draw the boundary between what we call Eocene and what we call Oligocene did change. Because of this new placement of the Eocene-Oligocene boundary, a part of what had been considered the early Oligocene was shifted into the late Eocene, and we therefore now refer to Florissant as late Eocene in age. And by the new radiometric dates, it was the *very end* of the Eocene Epoch.

This position in time—at the close of the Eocene—relates significantly to some of the aspects that make Florissant such an important fossil site. During this time, major changes in the world's climate were about to take place, and these impacted and changed the ancient Florissant environment forever. Florissant is a critical link in our understanding of the world just before these events happened. We turn now to the question of reconstructing this bygone world.

RELATED REFERENCES

Fossil preservation in diatom mats: Harding and Chant (2000); O'Brien et al. (1998, 2002)
Geology and stratigraphy of Florissant: Epis and Chapin (1974, 1975); Evanoff et al. (2001); MacGinitie (1953); McLeroy and Anderson (1966); Wobus and Epis (1978)

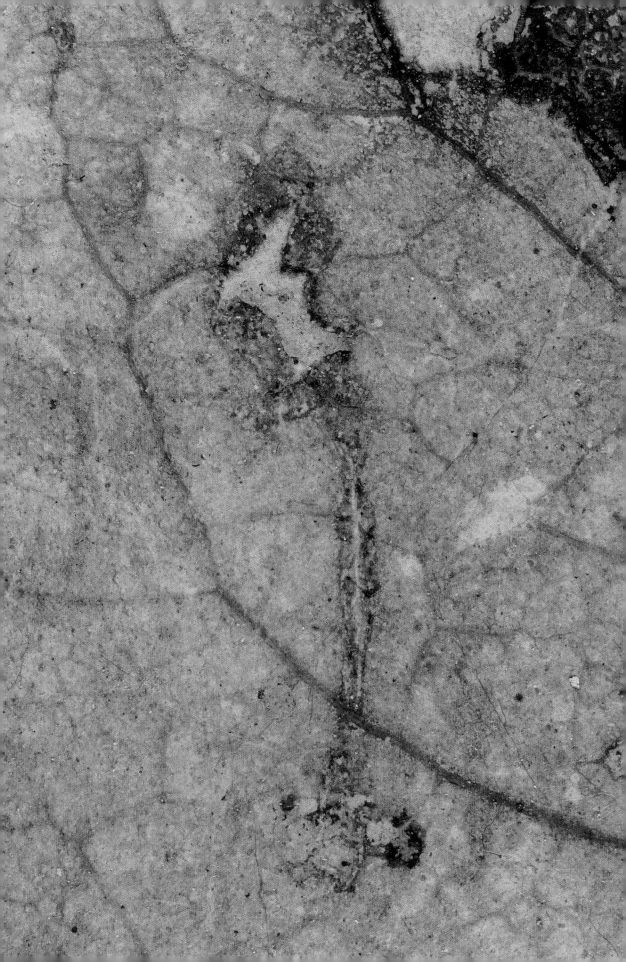

RECONSTRUCTING THE ANCIENT ECOSYSTEM

What makes Florissant significant? Many isolated facts contribute to Florissant's fame: a record of as many as 1,500 species of insects and 150 species of plants; more known species of fossil butterflies than any other fossil site; the only known fossil record of the tsetse fly; the only known petrified redwood trio; some of the largest diameter petrified trees in the world; and one of the world's earliest records of freshwater diatoms. Collectively, these various measures of high biotic diversity and uniqueness all contribute to Florissant's standing in the "world-renowned" category. Much more important, however, is the piece that Florissant reveals in the broader puzzle of climate change, the evolution of biotic communities, and mountain uplift in North America.

Primarily because of its tremendous insect diversity, Florissant ranks among the world's richest fossil sites as measured by the number of described species. Only a few other sites, such as the Baltic Amber from the Baltic Sea area, surpass Florissant for insect diversity. The record of insects and plants at Florissant is superb, but in spite of this diversity, pieces of the puzzle are still missing. Only a few mammals, and no reptiles or amphibians, have been found.

If you were able to travel back to the late Eocene and walk around ancient Lake Florissant, what would you find? The plants and insects would seem at least superficially familiar, with the majority belonging to genera that still live today. On closer examination, however, you would find some differences. You might pick up a leaf from a tree of the genus *Cedrelospermum* and, if you were familiar with modern trees, notice the close resemblance to the modern Caucasian elms (*Zelkova*), yet you would find that the winged fruits of *Cedrelospermum* were quite different. If you were a butter-

Left: Detail of leaf of *Trichilia* shows a rare example of leaf mining. From the collection of Florissant Fossil Beds National Monument.

fly collector, you might find *Prodryas,* an admiral-like nymphalid butterfly with wing patterns similar to those in certain modern nymphalids, but, unlike the closest modern relatives, *Prodryas* had a short tail on the wing. By the late Eocene, most of the plants and insects were similar, though not exactly identical, to those of today. By contrast, if you encountered one of the mammals, such as a brontothere, oreodont, or horse, you would find yourself facing an animal unlike anything in even the most exotic zoo. The fossils from Florissant help to demonstrate that different groups of organisms, such as plants and mammals—and different types of organs, such as leaves and flowers—have evolved at different rates.

Florissant's high diversity of both plants and insects provides important clues for understanding the evolution of biotic communities, including the interaction and coevolution of insects with their plant hosts. Considered individually, each of Florissant's species holds a wealth of information about evolution, geographic dispersal, and changing climatic tolerances. Considered collectively, all of Florissant's organisms reveal a much larger picture—one of an ancient community that thrived in a climate much different from that of Florissant today.

Florissant represents the period just preceding a very significant cooling of the world's climate during the transition from the Eocene to the Oligocene Epoch 34 to 33 million years ago. Florissant is the most diverse fossil deposit preserving an upland ecosystem immediately preceding this global cooling. It is a vital link in understanding the response of ancient North American plant and animal communities to this event, and it provides broader insight into the mechanisms and patterns of ecosystem evolution during periods of global climate change. It also holds evidence showing that the Rocky Mountains were already at high elevation by the late Eocene, which has significant implications for interpreting the timing and magnitude of mountain building in the southern Rocky Mountains.

RELATIVES NEAR, FAR, AND EXTINCT: FLORISSANT AS A UNIQUE BIOTIC COMMUNITY

As with most fossil deposits, the association of organisms at Florissant has no exact match in the modern world. All of the species, many of the genera, and even some of the families are extinct. Evolution has been more conservative, or slower, in some groups of organisms than in others. Most of the fossil plants and many of the insects can be placed into modern genera, whereas the few mammals from Florissant all represent extinct genera, and in the case of the brontothere and the oreodont, extinct families. The insects and spiders include about 280 genera that are described as extinct. Several of the fossil plant genera also are extinct, including the two most abundant plants at Florissant, *Cedrelospermum* and *Fagopsis,* extinct members of the elm and beech families.

Most of the fossil plants and insects, however, represent genera that still live today, yet many of their modern distributions do not overlap. These genera—once living side by side in the late Eocene forest at Florissant—are represented by related living species now restricted to widely separated parts of the world. These distributions attest to the existence of past corridors of biogeographic exchange. The disjunctions between past and present distributions of genera may be because organisms have dispersed into new areas, or because the modern occurrences are relicts of once more widespread distributions. Florissant includes many plants that are known, judging from other fossil sites, to have been widespread in a temperate forest that covered much of the Northern Hemisphere during the middle part of the Tertiary Period. Examples include *Sequoia* (redwood), *Acer* (maple), *Ailanthus* (tree of heaven), *Cedrelospermum, Eucommia* (hard-rubber tree), and *Hydrangea*. Some of them, such as *Sequoia, Ailanthus,* and *Eucommia,* are restricted in their modern ranges, whereas others, such as *Acer* and *Hydrangea,* are still widespread. *Cedrelospermum,* however, is now extinct.

Among the fossil plants from Florissant, some of the strongest affinities can be found with the modern floras of northeast Mexico and southern Texas, southeast Asia, the west coast of North America, the southern Rocky Mountains, and the southern Appalachian Mountains. There is, however, no single region in the modern world where all members of Florissant's ancient community still coexist. Today, *Sequoia* is unique to the California coast. *Koelreuteria* (golden-rain tree) and *Eucommia* are endemic to eastern Asia, whereas *Ailanthus* (tree of heaven) ranges from eastern and southern Asia into northern Australia. The vine *Thouinia* now grows in Mexico and the West Indies. *Glossina* (tsetse fly) lives exclusively in equatorial Africa. Relatives of *Dominickus* (butterfly moth) are restricted to Central and South America. *Rhingiopsis* (one of the soldier fly genera) lives only in the New World Tropics today. Other Florissant fossils represent genera that still inhabit the southern Rocky Mountains today, including plants such as *Pinus* (pine), *Picea* (spruce), *Salix* (willow), *Rosa* (rose), *Cercocarpus* (mountain mahogany), and *Ribes* (currant), and insects such as *Vanessa* (painted lady butterfly), *Bibio* (march fly), *Hydropsyche* (caddisfly), *Aphaenogaster* (ant), and *Harpalus* (ground beetle).

Another striking aspect of Florissant's ancient community is that it includes organisms whose modern relatives occupy not only different regions, but also different climatic habitats. Plants and insects that are today subtropical lived beside others that are typically temperate in modern distribution. MacGinitie considered that some of the fossil plants whose living relatives are tropical may have been species that were adapted to live in a temperate climate during the late Eocene. What has changed? In some cases, organisms evolved different adaptations to the environment through time, and hence their living relatives now inhabit different ecological niches. The ancient Florissant community shows just one snapshot in a long sequence of ecosystem evolution. Several responses of individual species—extinction, evolution, and dispersal to faraway places—converged to shape the fate of the community that once lived at Florissant.

The associations of species that make up biotic communities change in composition over time. This community evolution is a complex, dynamic process stimulated largely by physical events such as climate change. As aspects of the physical environment change, each individual species within a community has its own capacity for responding to these changes. During an episode of significant climatic cooling, for example, some species may become extinct whereas others have ranges of tolerance that allow them to exist under the more extreme conditions. Some species have the genetic capability to evolve rapidly and adapt to the new conditions. Other species may move by dispersal or migration to find niches in more favorable habitats. As new types of environments form because of climate change, organisms undergo adaptive radiations, diversifying by evolution into new species and genera. This complex combination of factors causes the association of species within a community to change. Because of these changes, species that live together today may have had ancestors that lived in dissimilar communities in the past. Likewise, species that lived together in the past—such as those at Florissant—may have descendants that occupy different communities in the modern world. For these reasons, the late Eocene biotic community at Florissant represents a congregation of species unique to that time and place.

EVIDENCE OF THE LATE EOCENE CLIMATE

Today's montane coniferous forest at Florissant stands in stark contrast to the lush forest revealed by the fossil plants and insects of the late Eocene. The modern forest is adapted to the harsh, cold climate of the high Rocky Mountains and is dominated by needle-leaved evergreen conifers such as pine, spruce, and Douglas fir, along with deciduous broad-leaved angiosperms such as aspen and several types of shrubs. The late Eocene forest at Florissant included some conspicuous conifers as well, but some, such as the giant redwoods and false cypresses, would seem strangely out of place in the modern forest. Most of the fossil plants were flowering trees and shrubs—a diverse mixture of mostly deciduous types along with some evergreens—only a few of which have relatives at Florissant today. Among the fossil insects, tsetse flies seem to be the most exotic, far removed from their modern relatives in equatorial Africa. Even the most casual observer, looking at an exhibit of Florissant's fossils and then walking through the modern forest, can see that the ancient environment here must have been very different.

Fossil plants, and fossil leaves in particular, have proven to be the most useful and reliable indicators for reconstructing ancient climate and environment. Plants are sensitive indicators of climate, and they are, in a sense, our thermometers of the past. Whereas many animals have the mobility that enables them to respond rapidly to seasonal or daily change by migrating to more hospitable environments, plants, on the other hand, must possess adaptations that allow them to survive the full range of environmental extremes in the location where they become established. These adaptations are both physiological and morphological. Although physiological aspects are not preserved in the fossils, it *is* possible to examine morphological characters that show strong correlation with climate. It is also possible to look at where the modern relatives of these fossils are living today. And from these clues, it is possible to reconstruct the temperature and precipitation of the past.

One method for determining climate from fossil plants relies on the present-day climatic distribution of their nearest living relatives. By analyzing this distribution, we can draw conclusions about paleoclimate. There are two problems with this method. First, it relies on correct identification of the fossils before valid comparisons with living species can be attempted, and it is not uncommon that fossils are originally misidentified and later reclassified. For example, several of the fossil plants originally placed into modern genera have later been shown actually to represent extinct genera. In such cases, using the distribution of the living genera would have been invalid and the conclusions about paleoclimate would be erroneous.

The second problem with the nearest-living-relative method is that, through evolution, organisms can adapt to different climatic conditions. Florissant fossil plants represent a mixture of types, some whose modern relatives are temperate and others, subtropical. The problem is that an organism living in a particular type of climate today may have lived in a different climate during the past. Even though a living species and a fossil species may have morphological similarities, their physiological tolerances to the environment may have changed. The principle of uniformitarianism, which is based on processes of physical geology and is simply stated as "the present is the key to the past," does not apply well to the biological world, especially where the climatic tolerances of individual organisms are concerned. Unlike most physical processes, all organisms have the potential to evolve and change. For that reason, the nearest-living-relative method must be used with caution. With some plants, such as palms, it can be used to infer climate with some degree of success, but with others it is more difficult. New methods are being developed to better measure the "coexistence," or climatic overlap, between the various nearest living relatives for fossil floras in hopes of proving the validity of this method.

The nearest-living-relative method also could be used with the fossil insects, which, like the plants, show a mixture of temperate and subtropical types at Florissant, but the same problems would apply. Modern tsetse flies, for example, live only in equatorial tropical climates where mean annual temperatures are 18–30°C. It is possible, however, that their ancestors may have lived under different conditions, and that only the modern representatives are so restricted by climate. As the entomologist Maurice James (1939, p. 48) noted of the Florissant fossil insects, "Though it is hard to explain the occurrence of five species of Nemestrinidae in a humid climate, it would be more difficult to explain an abundance of Bibionidae and Empididae . . . in a warm, arid region." Clearly, the nearest-living-relative method for estimating paleoclimate, whether used with plants or insects, presents contradictions.

Because of the problems inherent to the nearest-living-relative method, many paleobotanists of recent years have chosen to use another approach that is thought to give more reliable results in estimating paleoclimate. This second method relies upon the fact that the gross appearance, or physiognomy, of plant growth forms is dependent upon climate. Certain plant forms evolve as adaptations to particular conditions of temperature and precipitation. These include growth stature, number of canopy layers in a forest, whether or not there are vines dangling from the trees, and characters of the leaves such as size, shape, and the presence or absence of teeth on the margins. Other important aspects include the relative proportions of coniferous, broad-leaved evergreen, and broad-leaved deciduous plants in a flora. In the modern world, unrelated groups of plants living under similar climatic conditions share many physiognomic features as adaptations to the particular kind of climate in which they grow. For example, unrelated tropical plants from different regions of the world have evolved identical physiognomies because they live under the same conditions of climate. Just as these physiognomic characters are repeated in similar climates of different regions today, so, too, can these correlations be repeated through geologic time.

Leaves are one of the most commonly preserved plant organs in the fossil record, and they have physiognomic characteristics of shape and form that correlate closely with climate. One of the most useful features to note is whether the leaf margin is entire (i.e., smooth) or non-entire (i.e., toothed or lobed). Various studies have demonstrated that there is a strong correlation between the percentage of entire-margined species in a flora and mean annual temperature, with higher percentages of entire margins representing proportionately higher temperatures. Other leaf characters that show correlations with climate include size, texture, and the presence of long, tapering "drip-tips" at the apex. These features of fossil leaves make it possible to reconstruct paleoclimate, and particularly temperature, by using the correlations of these features with modern climate. Even if a fossil is not properly identified taxonomically, its physiognomic characters represent physical adaptations that still can be correlated with climate.

Both the nearest-living-relative and physiognomic methods have been used at Florissant to reconstruct paleoclimate, each giving different results. Either method can be used with the fossil leaves, but the fossil pollen and spores must be analyzed by using the nearest-living-relative method because there are no physiognomic characters evident in pollen and spores. The nearest-living-relative method, first applied to Florissant leaves and fruits by MacGinitie in 1953, gives the warmest results, with a mean annual temperature of about 16–18°C. By contrast, variations of the physiognomic method used in studies by Wolfe and Gregory during the 1990s give cooler results of about 11–13°C. The physiognomic results are probably less ambiguous, in view of the problems discussed above, and they can be more accurately and precisely quantified statistically because of the many documented measurements from modern forests. In my interpretation, based mostly on physiognomy but also considering evidence of nearest living relatives, Florissant's mean annual temperature was about 13°C, roughly the same temperature (but clearly not the same seasonality of precipitation) as the modern San Francisco Bay area. By comparison, the modern mean annual temperature at Florissant is a frigid 4°C. Another aspect of climate is equability, which is an expression of the seasonal variation in temperature. This can be measured as the difference between the mean temperatures of the warmest and coldest months. The evidence for Florissant's ancient climate shows that this difference was only moderate, indicating a comparatively equable climate lacking prolonged periods of severe cold or extreme heat.

Other evidence of climate comes from the fossil pollen and spores. At Florissant, these represent a mixture of mostly warm temperate types with other types whose nearest living relatives range into subtropical climates. Leopold and Clay-Poole, using the nearest-living-relative method for pollen, suggested that some of these could not live in a mean annual temperature lower than 17.5°C. Arguably, however, pollen is less diagnostic taxonomically than are leaves. It is also dispersed more widely by the wind and therefore represents a mixture from different environments; and again, the climatic tolerances of these modern relatives may have changed through time.

The discrepancy between these different temperature estimates serves as a good example of the scientific method, wherein hypotheses and results need to be retested as new methods are developed. Sometimes, definitive answers are difficult to prove, and old ideas die hard. As we will see, these estimates of paleotemperature are a critical step in estimating the paleoelevation at Florissant.

Estimates of the late Eocene precipitation at Florissant indicate about 50–80 centimeters of annual rainfall, with a distinct dry season. This moderate amount of precipitation does not indicate a particularly wet climate, but it was nevertheless wetter than Florissant's modern precipitation of 38 centimeters. Evidence for the ancient precipitation comes from the shape and relatively small size of the leaves, and features of the teeth, all characteristics that are correlated with seasonally dry climates. It is also indicated by some of the modern living relatives of the fossils, many of which inhabit areas with limited seasonal precipitation. Most of the rainfall fell during the growing season, in the late spring and early summer. Winter snowfall was probably rare except perhaps on the higher slopes of the Guffey volcano.

Growth rings preserved in the petrified trees give additional clues to Florissant's ancient climate. Cross-sections of the fossil redwoods reveal that the average ring width, an indicator of overall growth, is greater than in modern redwoods. This suggests that the late Eocene redwoods at Florissant grew under more favorable conditions than do modern redwoods in coastal California, perhaps because more precipitation fell during the growing season, or because the ancient trees grew along a valley bottom, where the soil was moister. The fossils also show a sharp transition between earlywood (consisting of large thin-walled cells formed early in the season) and latewood (consisting of small thick-walled cells formed late in the season). This transition indicates a rapid end to the growing season due to a lack of rainfall, lower temperatures, or shorter day length. Examination of the broad-leaved hardwoods shows that most of these are ring-porous; that is, they have distinct growth rings, indicating that the climate was seasonal with a definite dry or cold season during which growth was inhibited.

The ancient Florissant climate can be described in general terms as warm temperate (although definitions of this vary) with only modest rainfall, most of which came during the late spring and summer. A modern region with temperatures and precipitation similar to those estimated for Florissant can be found in the Sierra Madre ranges of north-central Mexico. Here, between about 1,500 and 2,500 meters, annual precipitation is generally 50–65 centimeters, with most of it falling during the summer months. Mean annual temperatures within the ranges estimated for Florissant (11–18°C) occur along this elevation gradient in the Mexican highlands. This is not to say that the ancient climate and vegetation at Florissant were identical to those of mountainous north-central Mexico, but the comparison is noteworthy. Indeed, workers such as MacGinitie and Leopold have noted the resemblance of some aspects of the Florissant plant community with modern communities in the Sierra Madre Orientale of northeastern Mexico, especially with the streamside communities.

MICROHABITATS

The ancient community at Florissant was not homogeneous across the landscape. Instead, it was a complex mosaic of microhabitat types that varied from one site to another. The most commonly fossilized plants are often those that grow nearest to streams and lakes, and for that reason, fossil floras can appear deceptively moist in character. More distant plants or those adapted to drier conditions are represented by fewer fossils. But by looking at the relative abundances of the different plants in the fossil record (a measurement that in part reflects their proximity to the sites of deposition), and at the preferred habitats of their modern relatives, it is possible to make inferences about the patterns of vegetation distribution across the ancient landscape. This evidence comes from leaves as well as pollen. Microhabitat differences—from the lake to the dry hillsides to the nearby uplands—indicate that there were steep ecological gradients in the region surrounding Lake Florissant.

Plants of the aquatic lake environment include algae, *Azolla* (free-floating water fern), and aquatic angiosperms such as Nymphaeaceae (water lilies), *Typha* (cattails), and *Potamogeton* (pondweed). These plants indicate a freshwater lake that was locally shallow and marshy near the shoreline. The terrestrial vegetation was lush along the lake and streams, where the local environment provided plentiful moisture, but the nearby hillsides and ridge tops supported dry-adapted vegetation. Valley-bottom sites alongside the lake were dominated by

Fagopsis, Cedrelospermum, Populus (poplar), and *Salix* (willow). The valley-bottom vegetation also consisted of groves of tall *Sequoia* and *Chamaecyparis* (white cedar or false cypress) trees. The understory of this tall valley-bottom forest consisted of smaller trees and shrubs including *Carya* (hickory), *Koelreuteria, Acer, Sapindus* (soapberry), and *Paracarpinus* (an extinct hornbeam-like genus). There was probably a gradual transition in the plant community between the valley and the surrounding hills, with shrubs such as *Amelanchier* (serviceberry), *Rosa, Rhus* (sumac), *Ribes* (currant), and *Staphylea* (bladdernut) extending into both habitats. The hillsides and ridges surrounding the lake basin supported open, drier woodlands and scrublands dominated by *Pinus* (pine), *Quercus* (oak), *Cercocarpus* (mountain mahogany), and *Vauquelinia* (rosewood). Still drier or even salty microhabitats are indicated by fossil pollen of the family Chenopodiaceae (saltbush or goosefoot family) and the genus *Ephedra* (Mormon tea), both typical of arid sites today. More distant uplands, perhaps including the slopes of the Guffey volcano, supported coniferous forests of pine, spruce, and fir (*Abies*). Spruce and fir might also have lived in locally cool places along the steeper shaded slopes above the stream drainages.

The insects at Florissant also showed patterns of microhabitat distribution. Insects of the lake ranged from exclusively aquatic groups such as water scavenger beetles to groups such as caddisflies and dragonflies, which lived near the water and completed a part of their life cycle aquatically. Most crane flies and midges lived in the damp habitats of thick vegetation near the lake and their larvae were aquatic, whereas many of the butterflies and bees, whose larvae were terrestrial, preferred open areas in the surrounding meadows and hillsides. Some of the Florissant fossil insects are known both as aquatic larvae and as flying adults, each occupying different habitats during their different life stages. Because insects have mobility, many species were wide ranging between varied habitats. Some species of insects lived in close association with particular plants, and their niches were determined by the distribution of those host plants.

THE IMPACT OF VOLCANIC ERUPTIONS

The microhabitats of plant communities were, at times, influenced and changed by the volcanic eruptions that spread ash and pumice across the landscape. The evidence for this is preserved in the record of fossil pollen and spores. Wingate and Nichols examined the change in pollen and spore frequencies through stratigraphic sequences of the lake shales to see how plant abundances changed through time. This information is presented as a pollen diagram (Figure 38).

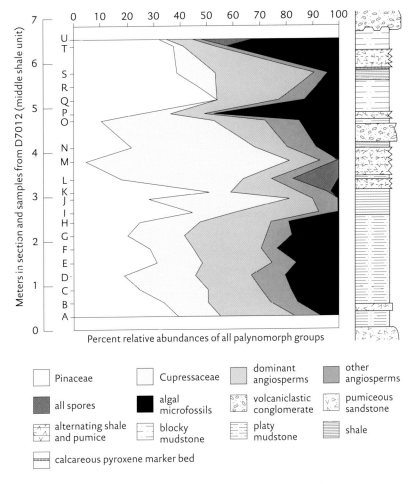

Figure 38. This pollen diagram shows the change in pollen and spore percentages through a section of the middle lake shales. These percentages reflect changes in plant abundances over relatively short intervals of geologic time. The diagram is based on pollen that was counted from samples taken from the shale at intervals about 20 to 50 cm apart. The most significant change occurs between three and five meters, where Pinaceae (pine, spruce, and fir) pollen decreases and Cupressaceae (redwood and false cypress) pollen and fern spores increase. This is interpreted as a change in the composition of the local vegetation in response to an interval of major volcanic eruptions, as indicated by the increased amounts of pumice and other volcanic sediment in the rock column (to the right) coincident with the change in pollen percentages in the pollen diagram. Diagram is from the work of Wingate and Nichols (2001). Figure from Evanoff et al. (2001). © 2001 by the Denver Museum of Nature & Science, used by permission.

Changes in plant abundances over geologically short intervals of time could be due to several factors, such as minor climate fluctuations, disturbance of the landscape by volcanic eruptions, developmental stages in ecological community succession, changes in lake level, or infilling of the lake. The pollen diagram shows a change in plant abundances in a section of the middle shale unit, with Pinaceae (pines, spruces, and firs) decreasing significantly relative to the increasing abundance of Cupressaceae (redwoods and false cypresses), as well as an increase in spores. This is interpreted as a change in the composition of the local vegetation in response to an interval of major volcanic eruptions, although fire also may have contributed to this change. The pines growing on the slopes evidently were reduced in percentage relative to the redwoods and white cedars growing closer to the lake, and the increase in spores suggests proliferation of ferns, often the first plants to recolonize a volcanically disturbed landscape. The correlation of this vegetation change with a major eruptive interval is indicated by the thick beds of pumice and other volcanic sediment in that portion of the stratigraphic section. There is, however, no evidence for frequent or repeated mass destruction of vegetation followed by cycles of ecological community succession. Repetitive minor ash falls are recorded throughout the lake sediments and apparently had no significant impact on the vegetation. Only major eruptions appear to have affected the local vegetation around Lake Florissant. For most of the time, the Guffey volcano sat quiet, and when it did erupt, the consequences were only occasionally catastrophic for the biotic community.

A WORLD OF WARMTH: THE EOCENE GREENHOUSE

The world's climate was warm throughout the Eocene Epoch. The early Eocene was quite warm, but there were notable fluctuations in temperature with a general trend toward cooling as the Eocene progressed. During the Eocene, lowland mean annual temperatures ranged from 14 to 27°C. The evidence for this comes from fossil plants, analyzed by Wolfe using the physiognomic method. Other evidence supporting these climate changes comes from oxygen isotope ratios ($^{18}O/^{16}O$) in the oceans, which fluctuate according to temperature and are preserved in the shells of fossil marine organisms. During the Eocene, subtropical forests extended into the Arctic. Late Eocene fossil floras from coastal areas of the Pacific Northwest and the Gulf Coast of the United States, comparable in age to Florissant, indicate that conditions in these lowland regions were much warmer than at Florissant. As we shall see, the cooler conditions indicated for Florissant at that time probably were due to its location in an upland area of the continental interior. But for an upland area, it was still warm.

ORIGIN OF THE FLORISSANT COMMUNITY

To understand Florissant's position in the broader scheme of community evolution and climate change, we need to look at other fossil sites both older and younger than Florissant. The story is a complex one—too intricate be covered in detail here—but some important comparisons will serve as illustrations.

We begin this comparison in the middle Eocene, about 50 million years ago, with the plants and insects from the lake-deposited Green River Formation of northwest Colorado and adjacent Utah and Wyoming. The Green River community thrived in a dry subtropical climate warmer than Florissant's, and it probably lived at a somewhat lower elevation. As one of the most complete inland fossil sites older than Florissant, Green River provides a critical point for comparison. Just as the Florissant Formation preserves a snapshot in time of the latest Eocene biota, the Green River Formation provides a glimpse of vegetation at intervals of the middle Eocene, some 16 million years preceding Florissant.

MacGinitie published on the Green River fossil plants in 1969 and noted the striking similarity to many of the much younger Florissant plants. He found that 29 of the Green River species (41 percent of the total) were closely related to those from Florissant. Examples from this long list include species in such genera as *Sequoia, Pinus, Populus, Quercus, Celtis* (hackberry), *Cedrelospermum, Aristolochia, Rosa, Vauquelinia, Ailanthus, Cedrela, Rhus, Cardiospermum, Koelreuteria,* and *Sapindus.* Despite the similarities, there are important differences as well, and the two floras were by no means identical. Green River, for example, includes a variety of tropical forms not found at Florissant, whereas Florissant has a much greater abundance of temperate genera.

The strong similarities between Green River and Florissant show that the two communities probably stem from a common origin, and in many respects, the Florissant community can be considered a derivative of the Green River community. The organisms at Green River, of course, had come together from still older communities. The comparison between these two communities is one example to illustrate how communities can change through time. The warm temperate community at Florissant was born from the congregation of organisms from a variety of older communities. Many ancestors of those organisms can be traced back to the subtropical community of Green River. The remainder, many of which were more temperate in character, such as members of the rose family, were probably integrated into the Florissant community from older upland sources.

OUT OF THE GREENHOUSE, INTO THE ICE HOUSE: GLOBAL CLIMATE CHANGE AND THE FLORISSANT COMMUNITY

Florissant represents an important time in the overall pattern of world climate change. Radiometric dates of 34.07 million years obtained for the Florissant Formation indicate that the fossils were deposited near the very close of the Eocene Epoch, which ended 33.8 million years ago.

During the transition from the Eocene to the Oligocene Epoch, between 34 and 33 million years ago, the world's terrestrial climate cooled significantly. The evidence from the Pacific Northwest, where many lowland fossil floras document this change, shows that the climatic deterioration transformed the environment there from warm subtropical to temperate over a geologically short span of about a million years. Mean annual temperature in the midlatitudes cooled by about 8–10°C, and winters, in particular, became much colder. A change of that magnitude would have significant biological impacts should it happen suddenly today, and indeed, the Eocene-Oligocene cooling event is one of the biggest—if not *the* biggest—biotic events of the last 65 million years, since the extinction of the dinosaurs.

What caused the climate to change so drastically between the late Eocene and early Oligocene? Several causes have been proposed, among them changes in oceanic circulation, an increase in world volcanism, extraterrestrial impacts, and changes in the earth's rotational axis. The most likely cause of this change was the separation of Antarctica, South America, and Australia resulting from plate tectonic processes. This, in turn, reconfigured the world's oceanic circulation patterns, including the development of a circum-Antarctic current. Glaciation had begun on Antarctica, and cold waters existed along the margin of that continent. These cold waters were carried into equatorial regions by newly developing ocean-bottom currents. Aspects of this change in the world's oceanic circulation is what is now thought to have provided the stimulus for the global climate cooling on land during the Eocene-Oligocene transition. With the coming of the Oligocene Epoch, global climate cooled significantly, never again to warm to the levels known during the Eocene. It was necessary for life to adjust, and in the process, the late Eocene Florissant community was forced to change into something new.

Many of the changes that happened at Florissant between the late Eocene and the present are not recorded in the fossil record, but two younger fossil floras from nearby regions provide insight into the pattern of climate change and community evolution following Florissant's time.

One of the significant early Oligocene floras showing affinities with Florissant is the Bridge Creek flora from the John Day Formation in Oregon, which grew in a region of much lower elevation. It occurs in lake deposits 31.8–33.6 million years old, about 1–2 million years younger than Florissant. Insect fossils are extremely rare there. The fossil plants represent a temperate broad-leaved deciduous forest similar to the Mixed Mesophytic Forest growing in China today. The mean annual temperature was about 9–11°C, approximately 3°C cooler than Florissant, but the summers were warmer and the winters cooler. As the global climatic cooling opened new ecological niches, and new opportunities for community evolution, the Bridge Creek flora replaced the much warmer subtropical forests that grew in the Pacific Northwest lowlands during the late Eocene. Many of these subtropical plants became extinct, or they were regionally extirpated during the climatic change and their relicts survived only in the warmer climates of lower latitude. As the climate cooled during the early Oligocene and new plant communities came into existence, many genera dispersed from late Eocene upland communities such as Florissant into the cooled lowland communities, such as Bridge Creek (Figure 39). Many of the genera found at Florissant also occur at Bridge Creek, including *Torreya* ("nutmeg" or "stinking yew"), *Abies, Pinus, Sequoia, Mahonia* (Oregon grape), *Quercus, Asterocarpinus* and *Paracarpinus* (extinct genera), *Carya, Juglans* (walnut), *Florissantia* (extinct genus), *Cedrelospermum, Ulmus* (elm), *Ribes, Amelanchier, Crataegus* (hawthorn), *Malus* (apple), *Rosa, Rubus* (blackberry), and *Acer,* among others. Despite these shared genera, both Florissant and Bridge Creek contain many other genera that are not shared, and indeed, they represent substantially different plant communities and ecosystems. The comparison of the two floras, however, provides an excellent illustration of how plant communities change dynamically as a response to climatic stimuli.

Other Florissant genera appear to have survived in regions to the south. Some of the more dry-adapted Florissant plants such as members of the families Leguminosae (legumes) and Anacardiaceae (sumacs), and insects such as *Dominickus* (butterfly moth) and *Rhingiopsis* (one of the soldier fly genera), probably dispersed to, or persisted in, the refugia of lower elevations at lower latitudes, where some of their descendents remain today.

A second significant fossil site is at Creede, Colorado, located about 190 kilometers southwest of Florissant. Creede is late Oligocene in age, dated at 27.2 million years, 7 million years younger than Florissant. The plant and insect fossils at this site are preserved in lake deposits that formed within a caldera. Creede represents a cool temperate coniferous forest with fir, spruce, and pine inter-

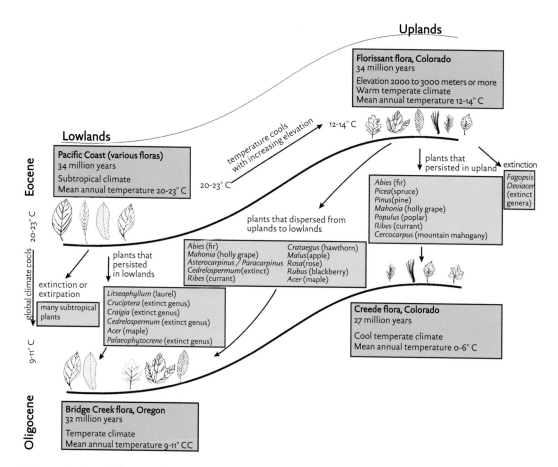

Figure 39. This model illustrates the dynamics of changing biotic communities during the major climatic deterioration of the Eocene-Oligocene transition. The top profile idealizes a transect from the Pacific Coast to the upland interior of the Rocky Mountains during the latest part of the Eocene Epoch 33–34 million years ago, and the bottom profile does the same for the Oligocene Epoch 27–33 million years ago. The lowland of the Eocene was subtropical, whereas the upland at Florissant was warm temperate owing to its higher elevation. This ancient elevation can be estimated from the temperatures indicated from fossil plants, because temperatures change at a constant rate with increasing elevation. Following the drastic climatic cooling, the lowland of the early Oligocene became warm temperate, and the upland in the Rocky Mountains became cool temperate. During this climate change, whole communities of plants and animals did not move as intact units, but rather, each individual species responded in one of several ways. Some species became extinct or were extirpated (eliminated) regionally, others persisted in the same region because they possessed or evolved adaptations to tolerate cooler conditions, and still others dispersed from upland regions to lowland regions in order to remain in approximately the same type of climate. Some of the plant genera that continued into the Oligocene became extinct at a later time. The genera illustrated in this diagram are selected examples from longer lists for each category, and in some cases, a genus can appear as an alternative in more than one category. The illustrations of fossil plants in the diagram show that characteristics of leaf size and margin are different in different climates, and this is one method by which paleobotanists reconstruct ancient environments. Drafted by Linda Lutz-Ryan.

spersed with woodlands of pine and juniper (*Juniperus*) and chaparral of mountain mahogany (*Cercocarpus*). Other smaller shrubs of the understory included Oregon grape and currant. The climate was much cooler than it had been at Florissant during the late Eocene, with mean annual temperature estimated as 0–6°C. A direct comparison with Florissant is complicated by possible differences in elevation, and Creede is estimated to have been up to several hundred meters lower. Summers were dry and much of the region's precipitation came as snow. Compared with Florissant, Creede has a higher proportion of its plant genera in common with the present flora of the southern Rocky Mountains, along with more physiognomic similarities to the modern vegetation of the region, indicating that a definite trend toward modernization of the flora and vegetation in this region was underway by the late Oligocene. Some of the Creede genera, such as *Abies, Picea, Pinus, Mahonia, Populus, Ribes,* and *Cercocarpus,* had been present in the late Eocene at Florissant. Either these genera were capable of living in a cool climate even as early as the late Eocene, or else they evolved this capability in response to the cooling climate. They continue to be important genera in the modern montane communities of the Rocky Mountains. Compared with Florissant, Creede shows that the biotic communities of the Rocky Mountains were one step closer to those of today. Closer, but they were still different.

As environments and climates change, biotic communities seldom move as intact groups to new locations. Instead, each individual species has its own genetic capabilities which ultimately determine its survival, evolution, extinction, and rate of dispersal. Some members of the ancient Florissant community, such as the very common beechlike *Fagopsis* and the huge brontotheres, were soon to become extinct at the close of the Eocene. Many others were extirpated from the Florissant area as they were geographically reshuffled in response to the changing distribution of environments, finding new niches through dispersal and evolutionary adaptation in a variety of newly formed and ever-changing communities. Many of the insects, because of their mobility, probably were able to migrate quickly to more favorable geographic regions, whereas plants dispersed at slower rates. Hardy, cold-tolerant members of the Florissant community continued to survive in the region. Over longer intervals of geologic time, new species evolved from those that once inhabited Florissant. Through a combination of these factors, the ancient Florissant community became forever disaggregated.

PALEOELEVATION: AN EOCENE ROCKY MOUNTAIN HIGH?

The elevation at Florissant during the late Eocene has been a much debated topic recently. MacGinitie, in his 1953 monograph, was the first to use the fossil plants for estimating paleoelevation. Comparing the fossils with the distribution of their living relatives, he concluded that the Florissant forest grew at an elevation of about 300 to 900 meters. This is much lower than Florissant's modern elevation of 2,500–2,600 meters in the valley where the fossils occur. For four decades, MacGinitie's estimate was the benchmark for the late Eocene elevation of this region, and it was cited in geologic studies as evidence that the region must have been significantly uplifted during the later part of the Tertiary Period. This estimate played a key role in interpreting the history of uplift and mountain building in the southern Rocky Mountains.

During the 1990s, new studies conducted independently by Gregory, Wolfe, and me all pointed to a different conclusion. New methods were developed using estimates of paleotemperature (as discussed previously) and lapse rates (the rate at which temperature decreases with increasing elevation) to measure paleoelevation. Most simply, paleoelevation is calculated first by using fossil plants to estimate two paleotemperatures: that at Florissant, and that from another fossil flora of the same age and at the same latitude at sea level. Modern lapse rates range from 3.0 to 8.0°C per 1,000 meters, or, put differently, mean temperature decreases 1°C every time you ascend 125 to 333 meters in elevation. This variability exists because lapse rates differ from area to area, and because they can be calculated either over long distances or within particular areas. Paleoelevation can be calculated by taking the difference in paleotemperature between sea level and Florissant, and then multiplying that difference by the *appropriate* rate of elevation change for 1°C. This is a simplification of what is actually a much more complicated calculation, and reliable estimates of paleoelevation must consider many other complexities. For example, even at the same altitude, the temperature can vary between the coast and the continental interior, and large plateaus of high elevation can cause lapse rates to vary. In addition, fluctuation in sea level, changing climatic conditions, and paleogeography are all factors that enter into the calculation.

Another recent method proposed by Forest and others uses principles of atmospheric energy conservation to determine paleoelevation from estimates of paleoenthalpy (a thermodynamic property of the atmosphere, which can be estimated from fossil plants) rather than paleotemperature and lapse rate. Compared with the lapse rate method, this method better incorporates the effects of moisture on temperature distribution.

The consistent conclusion from these recent estimates is that Florissant was at a much higher paleoelevation than MacGinitie originally estimated. The estimates range from 1,900 to more than 4,100 meters, suggesting that during the late Eocene, Florissant was nearly as high as, or perhaps much higher than, the 2,500–2,600 meters that the valley is today. These interpretations suggest that global climate change, rather than uplift, has influenced the sharp contrast between the ancient and modern ecosystems at Florissant.

At least eleven studies have estimated Florissant's paleoelevation, most of them done during the 1990s. Not all of them agree, and even among those that conclude that the paleoelevation must have been high, the estimates differ, reflecting the paleotemperatures that are used and the way in which lapse rates are calculated, among other variables. Was Florissant already at its present elevation by the late Eocene? The recent paleobotanical estimates argue that it was, possibly even exceeding the present elevation and subsequently subsiding. Other paleobotanists and geologists remain skeptical, yet nevertheless, these high paleoelevation estimates present new challenges—to verify or refute the results—and perhaps to rewrite interpretations about the timing and forces of uplift that formed the southern Rocky Mountains.

PLANT-INSECT INTERACTIONS: THE LITTLE HOLES THAT TELL A BIG STORY

Although plants are the most useful indicators for reconstructing regional paleoclimate and large-scale community ecology, another realm of paleoecology and community evolution is revealed by the study of plant-insect interactions. Plants create many of the microenvironments that insects use for habitation and feeding. Ecological studies on the role of insects in the ancient environment therefore concentrate more on the relationship between insects and their plant hosts, rather than on large-scale climate reconstruction. These studies provide valuable information about the evolution of herbivorous feeding strategies, the coevolution between plants and insects, the development of ecosystem complexity, and the evolution of feeding patterns in response to climate change.

Much of the record of plant-insect interactions consists of trace fossils on leaves. These leaves show features of insect damage such as margin feeding, hole feeding, leaf mining, skeleton feeding, and galls (Figure 40). The presence of thickened reaction tissue around the wound (Figure 41) serves to distinguish

1. margin feeding
2. hole feeding
3. leaf mining
4. skeleton feeding
5. galls
6. leaf cutting

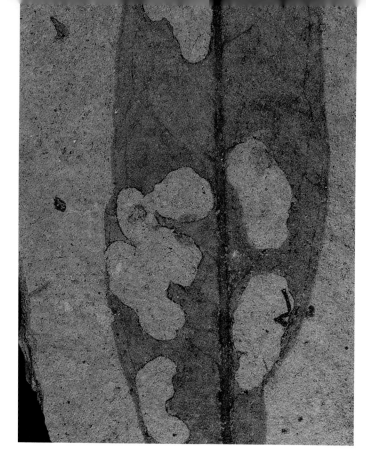

Figure 40. Plant-insect interactions can be classified according to the type of feeding trace or gall that the insect leaves on a leaf. Feeding traces may be along the margin of the leaf or as holes within the leaf. Skeleton feeding occurs where the insect has selectively eaten the tissue but left the intervening veins. Galls are produced by the plant in response to certain insects that live on the leaves. Leaf mines are made by larvae that feed on tissues within the leaf, forming a sort of tunnel bounded by the layers of leaf cuticle on the outer surfaces. Leaf cutting involves removal of leaf pieces for purposes such as nest building. Diagram drawn by Linda Lutz-Ryan.

Figure 41. Besides the abundance of body fossils, insects also left their mark on the fossil record as feeding traces. This leaf of *"Eugenia"* shows extensive hole feeding by an unknown insect. The dark, thickened reaction tissue around the margins of the holes could have formed only before the leaf was detached from the plant, and not as physical deterioration during transport. Specimen UCMP-198427, courtesy of the University of California Museum of Paleontology.

Scale: × 4.3

RECONSTRUCTING THE ANCIENT ECOSYSTEM   65

feeding traces from abrasion damage that occurred during transport and deposition. The reaction tissue confirms that the damage occurred while the leaf was still alive, and before it was shed.

In a quantitative study of insect damage on Florissant fossil leaves by Smith, 23 percent of the leaves showed insect damage, with 1.4 percent of the total leaf area of all leaves having been removed by insect herbivory. This is much less than in six modern forests that have been sampled, where 72–90 percent of the leaves have been damaged and 4–10 percent of the leaf area has been removed. Although there may be biases against the preservation of insect-damaged leaves in the fossil record—such as more rapid decay of damaged leaves and the over-

Figure 42. This leaf of *Paracarpinus* shows deeply incised margin feeding that extends to the midrib. The insect that caused this damage consumed large areas of leaf tissue between the secondary veins. Plants often respond to insect herbivory by producing toxic compounds, and such reactions are important defenses in the coevolution of plants and insects. If present, these toxins become more concentrated in the leaf's veins, and probably for that reason, the insects that chewed this leaf avoided the secondary veins. Specimen UCMP-198423, courtesy of the University of California Museum of Paleontology.

Scale: × 5.4

 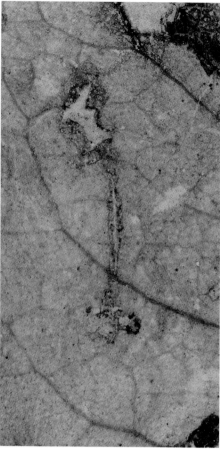

Figure 43. The pouchlike galls along the midrib of this leaf are thought to have formed in response to a gall mite (Class Arachnida, Order Acari, Family Eriophyidae). Several types of insects also could have been responsible for these galls, however, and it remains uncertain which type of mite or insect actually produced the damage on this leaf. The plant responded to the presence of these insects or mites by developing the galls. Galls such as these typically are located adjacent to the major veins of the leaf, where the plant's nutritive tissues are most concentrated. UCM-4429 (holotype), courtesy of the University of Colorado Museum.

Scale: × 7.7

Figure 44. This leaf of *Trichilia* shows a rare example of leaf mining at Florissant. The egg was laid at the base, and when the larva hatched, it formed the leaf mine as it tunneled through tissues within the leaf. The larva then made the pupation chamber, which is seen as the hole near the top. The larva developed into the pupa and later emerged as an adult. Leaf mines such as this are formed by Diptera (flies), Hymenoptera (bees and wasps), and Lepidoptera (butterflies and moths). Specimen FLFO-3514, from the collection of Florissant Fossil Beds National Monument. Interpretation courtesy of Dena Smith.

Scale: × 5.9

 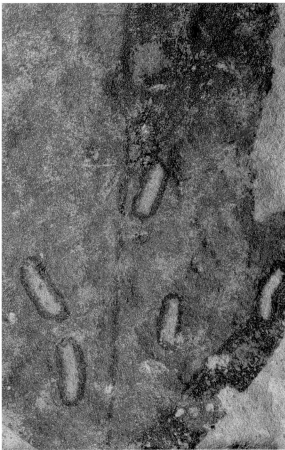

Figure 45. Leaf-cutting bees (Family Megachilidae) cut neat, circular pieces from leaves. The pieces are used to line the cells of the bee's nest. Most of the modern leaf-cutting bees are solitary and make their nests in natural cavities within wood, or sometimes in the ground. UCM-4543, courtesy of the University of Colorado Museum.

Scale: × 3.4

Figure 46. Leaves identified as *Cardiospermum* show a distinct type of insect damage specific only to the leaves of this plant. This unique type of hole-feeding damage probably was formed by the larvae of an insect species with specialized feeding habits. Specimen USNM-1883, courtesy of the National Museum of Natural History.

Scale: × 7.4

representation of small leaves on which insects feed less frequently—it nevertheless appears that the intensity of insect herbivory has increased significantly since the late Eocene.

Of the insect-damaged Florissant leaves, 88 percent were attacked by only one feeding group, whereas modern samples show that 41–67 percent of the leaves are attacked by multiple feeding groups. Most of the insect damage is hole feeding and margin feeding (Figures 41 and 42), whereas skeleton feeding and galling are much less frequent (Figure 43). Leaf mining is rare on Florissant leaves (Figure 44). Examples of leaf cutting, where portions of leaves have been removed for nest construction, are also present (Figure 45). Thirteen types of specialized feeding patterns occur, including types that are specific to *Cedrelospermum*, *Staphylea*, and *Cardiospermum* (Figure 46). Florissant leaves show no more than three feeding types on the same leaf, whereas many modern leaves often show more than three. All of these trends appear to show an increasing community complexity between plants and insects since the late Eocene—a trend that is evidently keeping herbivorous insects well fed

It has been suggested that insect herbivory is related to aspects of climate. For example, the degree of herbivory appears to increase with increasing temperature. Warmer sites at low latitude or altitude show greater insect herbivory than cooler sites at higher latitude or altitude. The degree of galling shows some correlation with precipitation, with a greater diversity of galls present in drier environments. Concentrations of atmospheric carbon dioxide (CO_2) also may influence insect herbivory, although in part this may result from the influence of CO_2 on temperature.

Much more work remains to be done in order to understand the complex coevolution between plants and insects. This coevolution is driven in part by adaptive radiations—that is, periods during which particular groups of plants and insects undergo rapid evolutionary diversification as they enter new habitats or develop new evolutionary strategies. A coevolutionary "arms race" may develop as certain insects evolve new mechanisms for exploiting plants, and the plants in turn evolve the defensive means to inhibit such exploitation. The study of insect-damaged fossils as a basis for understanding coevolution is a relatively new field in paleontology. Such studies use two different approaches. One involves looking at entire communities by examining all of the insect-damaged leaves from a fossil flora. The other approach looks at particular taxonomic groups, focusing, for example, on the insect damage to a specific group of plants through time. Sites such as Florissant—rich in both plants and insects—hold important clues for understanding these patterns of evolutionary ecology and community evolution.

THE FILTERS OF TAPHONOMY: BIASES IN THE FOSSIL RECORD

Paleontologists can never assume that the fossil record provides a completely realistic picture of the ancient plant and animal community. Many things can determine which organisms, or parts of organisms, enter the fossil record, and which do not. Taphonomy is the branch of paleontology that studies all of the factors in the transformation from a living community to a collection of fossils representing that community. These factors include characteristics of the organism or organ (e.g., the way in which leaves or fruits are shed); processes of death, decay, mechanical deterioration, sorting, transport by wind or water, settling, and deposition; post-depositional compaction and chemical changes; weathering and erosion of the rock; and biases in what the paleontologist chooses to collect. Whatever ends up in a museum collection only partially represents the reality of the ancient community, and any reconstruction of the community needs to consider these biases carefully. To understand the factors of taphonomy, paleontologists must study similar processes in modern environments.

Studies of modern environments show that the relative abundances of plants represented in sediment do not accurately reflect their relative abundances in the surrounding forest. Most of the plant parts deposited in a lake come from plants that grow in the immediate vicinity, and fewer come from more distant plants. Plants growing in the wet community along a stream or by a lake are easily overrepresented in the fossil record because their leaves and fruits can fall directly into the depositional basin. By contrast, plants living on the surrounding slopes and ridge tops are underrepresented because their parts must be transported over longer distances, if they are transported at all. Tall trees that shed leaves into a depositional basin are more likely to become fossilized than small herbaceous plants, which often die and wither in place. Deciduous trees and shrubs are better represented as fossils because they shed more leaves than evergreens, and plants with winged fruits or seeds easily transported by wind into a depositional basin are better known than those lacking this mechanism of dispersal. Thick-textured leaves are better able to survive the rigors of transport than thinner, more fragile leaves. Pollen from wind-pollinated plants accumulates like dust on a lake surface and is well represented in the fossil record, whereas pollen of insect-pollinated plants is carried directly from one flower to another with much less chance of entering a lake. For these reasons, some plants at Florissant are represented only by leaves, some only by fruits, and others only by pollen.

Less is known about the taphonomy of insects. In studies of a modern lake deposit, small ground-dwelling insects that eat plants or dead organisms were most commonly represented in the sediments, but aquatic and wood-inhabiting groups of insects were underrepresented. Only 28 percent of the beetle genera found living around the modern lake were actually preserved in the sediment, and of those, the relative abundances were quite different between the living and dead assemblages. The underrepresentation of aquatic insects in the lake sediment is curious, but may result from their softer bodies, which are less likely to become fossilized. Insects with hard exoskeletons, such as beetles, have a higher fossilization potential than softer insects such as aphids or caterpillars. At Florissant, winged ants are overrepresented relative to wingless ants. This may be because ants fly only during mating swarms, during which time they probably were carried over the lake where they could enter the depositional environment more easily than the wingless terrestrial forms.

Processes of deposition also influence what becomes fossilized. At Florissant, for example, lahars were ideal for preserving large tree trunks in their original position, but not leaves and insects. By contrast, diatom mats floating in the lake were influential in preserving the leaves and delicate insects that accumulated on the lake's surface. Postdepositional processes were also important aspects of taphonomy and included, for example, the precipitation of silica that caused the tree stumps to petrify. Collecting bias is another taphonomic factor in some of the large collections made by early paleontologists at Florissant, and it was not unknown for paleoentomologists to ignore the plants, or for paleobotanists to disregard the insects. Despite the huge taxonomic diversity of Florissant's fossil record, undoubtedly hundreds and probably thousands of other species were not fossilized because of the various aspects of taphonomy.

All reconstructions of the ancient ecosystem—community, climate, microhabitats, elevation, and plant-insect evolution—need to be tempered against the realities of taphonomy. What we get as fossils are only pieces—carefully selected by natural processes—of the complete yet never fully assembled picture.

RELATED REFERENCES

Biogeography: Manchester (1999)

Plants and paleoclimate: Gregory-Wodzicki (2001); MacGinitie (1953); Mossbrugger (1999); Wolfe (1978, 1992, 1994, 1995); Wolfe and Spicer (1999)

Florissant paleoclimate: Axelrod (1997); Gregory (1994); Gregory and Chase (1992); Leopold and Clay-Poole (2001); MacGinitie (1953); Meyer (1992); Wolfe (1992, 1994, 1995)

Wood growth rings: Gregory (2001); Wheeler (2001)

Pollen evidence of volcanic disturbance: Wingate and Nichols (2001)

Eocene-Oligocene transition: Prothero (1994); Wolfe (1992)

Green River, Bridge Creek, and Creede floras: Axelrod (1987); MacGinitie (1969); Meyer and Manchester (1997); Wolfe and Schorn (1990)

Evolution of plant communities: Mason (1947)

Paleoelevation: Axelrod (1997); Forest et al. (1995); Gregory (1994); Gregory and Chase (1992, 1994); MacGinitie 1953; Meyer (1992, 2001); Wolfe et al. (1998)

Plant-insect interactions: Scott and Titchener (1999); Smith (1998, 2000a)

Taphonomy: MacGinitie (1953); Smith (2000a, 2000b)

History of North American vegetation: Graham (1999)

FOSSIL PLANTS

From giant redwood stumps to microscopic pollen grains, Florissant's fossil plants span a wide range of sizes and include a variety of different organs. The petrified stumps, among the world's largest, are preserved upright in their original growth positions. Impressions and compressions of leaves, fruits, flowers, and cones are revealed as the pieces of shale split along their natural bedding planes. The shale also can be dissolved in various acids, leaving a microscopic residue of resistant pollen and spores. Each of these organs provides a different insight into the late Eocene flora of the southern Rocky Mountains.

The assortment of different plant organs presents paleobotanists with an unusual problem when trying to reconstruct extinct plants from fossils. This is because it is uncommon to find the different parts of a plant attached to one another in the fossil record. Instead, each type of organ is shed from the plant separately and transported independently into the depositional basin. Consequently, the leaves, fruits, wood, and pollen usually are preserved as isolated, unattached organs, making it difficult for paleobotanists to reconstruct which structures came from the same plant species. When these isolated organs are given taxonomic names, it is not always possible to demonstrate confidently that they belong to the same species, and for that reason a separate name sometimes must be given to each different organ. Organs such as pollen and wood may be distinguishable only to the generic or even family level, leaving doubt as to which of several leaf species might correspond to a particular type of pollen or wood.

Botanists who study modern plants base their classifications primarily upon reproductive characteristics of flowers and fruits, yet

Left: Detail of the cone scales of a female cone of *Pinus florissantii*. Courtesy of the National Museum of Natural History.

paleobotanists often need to make identifications and distinguish species only from leaves and wood. If various different organs—pollen, wood, leaves, and fruits—of a particular genus can be recognized, then the presence of that genus at Florissant can be more confidently verified. In rare instances fossil twigs show the actual attachment of foliage with reproductive structures, and these are important fossils for providing multiple-organ reconstructions of extinct species or genera that show more of the plant in its entirety. Such reconstructions sometimes reveal patterns of mosaic evolution, in which certain plant organs, such as fruits, evolved at more rapid rates, whereas others, such as leaves and wood, were more conservative in their evolutionary pace.

Paleobotanists base their identifications of fossil plants on careful comparison with modern plants as well as with other fossils. This usually requires a visit to a herbarium, where collections of pressed modern plants from different regions of the world can be studied. Valid identifications of fossil plants need to be based on diagnostic characteristics that are unique to a particular genus, but such features are not always well preserved in the fossil record. For example, three-dimensional flowers become distorted as they are fossilized in shale. The finest details of leaves, such as the smaller veins, frequently are difficult to see in the Florissant fossils. Adding to the difficulty, some unrelated groups of plants have evolved similar leaf features through convergent evolution, and it is only when multiple characteristics can be observed that convincing identifications can be made. For these reasons, some of the Florissant leaf identifications remain tentative.

Common names often have little meaning in botany, and for that reason, generic names always will be emphasized in our discussion of the fossil plants. Where common names are given, they are often only selected examples from a list of many common names that can be applied to a particular genus. In some cases, different genera have the same common name, such as the many distantly related "cedars."

As we have already seen, fossil plants are important in providing the clues for reconstructing past climate and elevation. They also provide important evidence for past corridors of intercontinental biotic exchange. For example, plant genera that are shared between North America and Europe or Asia help to answer questions about where these plants first originated, when they subsequently dispersed across land connections between these continents, and what ancient climatic and geographic conditions permitted this dispersal.

In this chapter, we will examine what can be learned from each major plant organ—wood, leaves, fruits, and pollen—in order to form a composite picture of the Florissant forest. Although all of these fossil plants represent extinct species, most, but not all, can be placed into modern genera. Among the extinct genera, however, are the two most common plants at Florissant: *Fagopsis* and *Cedrelospermum*.

FLORISSANT'S PETRIFIED FOREST

The petrified forest is one of the main attractions at Florissant Fossil Beds National Monument. It is difficult to know with certainty how many petrified trees it once contained, because so many of these were removed by overly enthusiastic collectors more than a century ago. A. C. Peale of the Hayden Survey mentioned seeing 20–30 stumps in the 1870s. More recent records document at

Figure 47. Nature continually recycles itself, sometimes in unusual ways. This ponderosa pine tree is rooted in the broken, weathered remains of a petrified redwood stump, slowly causing the fossil to disintegrate.

Figure 48. The Big Stump, although arguably not the "biggest" at Florissant, is a redwood (*Sequoia*) that measures 3.7 meters in diameter at breast height. This diameter is much greater at ground level, where the stump broadens near its base to a circumference of 18.5 meters and a diameter of 5.9 meters. This flared base, near the transition of the trunk into the shallow, spreading root system, helped to support the tall stature of the tree. Photograph by Lee Snapp.

Figure 49. The Redwood Trio has three separate but interconnected trunks growing together as one plant—a vegetative clone! Such clones develop in modern redwoods when the burls around the base of a stump or broken tree give rise to sprouts. The trio originated as three sprouts that grew into mature trunks, and the original trunk died and decayed. This type of reproductive growth form is unusual among conifers. The Florissant trio is the only known example of this in the fossil record. Photograph by Lee Snapp.

Figure 50. This cluster of trees in the modern redwood forest of California illustrates the same type of growth pattern preserved by the Redwood Trio at Florissant. Paleobotanist H. D. MacGinitie is standing near the base, ca. 1930s. Historic photograph from the collection of Florissant Fossil Beds National Monument.

least 30, some of which have been reburied for protection. Some of the stumps originally reported by Peale were later collected, whereas others were uncovered during the 1900s. Still others probably remain buried and undiscovered.

When they were living, these trees formed a grove along the valley bottom, with a forest canopy at least 60 meters high. A lahar, or debris flow, from the Guffey volcano carried a torrent of volcanic mud through the forest understory and entombed the bases of the trees in five meters of sediment, preserving them in situ. The trees died when the roots could no longer receive sufficient oxygen. Groundwater containing dissolved silica permeated the wood, precipitating silica within cell walls and petrifying the trees through the process of permineralization. This process preserved cellular details of the wood's anatomy and structure, enabling paleobotanists to identify the woods and to examine growth rings for information about paleoclimate. Although wood is prevalent, no bark has been identified from any of the Florissant trees.

The majority of the tree stumps identified at Florissant are similar to modern *Sequoia* (redwood), and several spectacular examples can be seen in the petrified forest area (Figures 47–49). The size of these giants is phenomenal—the Big Stump (Figure 48) measures 3.7 meters in diameter at breast height, and another stump measures more than 4.1 meters. They are among the largest-diameter petrified trees known. By comparison, modern redwoods typically measure 3–5 meters, and up to 10 meters, in diameter, and typically 65–100 meters in height. The Redwood Trio is a rarity for the fossil record in that it has three separate trunks that developed by sprouting (Figure 49). This type of reproductive habit also is seen in the modern redwood forests of California (Figure 50).

On the basis of its anatomy, the wood of the Big Stump was placed in the fossil genus *Sequoioxylon,* indicating that the wood shows a close relationship to the modern genus *Sequoia* in the family Taxodiaceae (Figure 51). It is the only type of coniferous wood that has been identified from Florissant. Minor differences in the structure of this *Sequoia*-like wood, however, indicate that it is not identical to the single living species of *Sequoia, S. sempervirens* (coast redwood). The bases of some of the large stumps are horizontally oriented, possibly showing the transition into roots. Study of the wood's growth rings provides evidence about growing conditions in the ancient environment, indicating that the Florissant trees grew under more favorable conditions than the modern redwoods. A comparison of sequences of growth rings between two of the stumps (a method known as cross-dating) shows a 180-year overlap, proving that these two trees, and probably all of the trees in the petrified forest, were growing at the same time. The growth rings indicate that the Redwood Trio was 500 to 700 years at the time of its death.

B

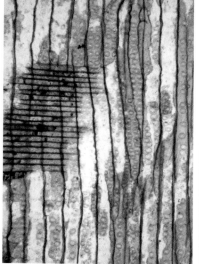

C

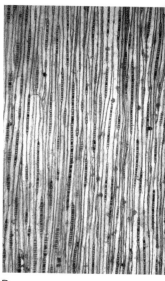

D

Figure 51. This *Sequoia*-like petrified wood has excellent cellular detail. To identify woods, it is necessary to view microscopic thin sections cut along three different planes (A). Each view shows different features of the wood's anatomy. A cross-section (B) is taken along a horizontal plane perpendicular to the axis of the stem's growth, and it shows growth rings as well as rays that look like lines crossing the growth rings at right angles. The open cells are the tracheids, which supported the tree and conducted fluids upward. The radial plane (C) is cut perpendicular to the cross-sectional plane and parallel to the axis of growth, and it passes from the outer surface of the stem toward the center of the stem, down along a radius of the tree trunk. The walls of the tracheids are visible as parallel vertical lines. A portion of a ray is visible left of center (the ray cell walls look like horizontal lines in this photograph). The circular structures in the tracheids were passageways for fluid exchange between the tracheids. The tangential plane (D) also passes through the stem parallel to the axis of growth, but it does so along a tangent to the growth rings, rather than at right angles through growth rings. The tangential view shows the tracheids (the long open areas) as well as scattered rays (the narrow vertical lines with tiered cells). Approximate dimensions of the photographs are 2.3 x 3.0 mm in A; 1.0 x 1.3 mm in B; and 2.3 x 3.1 mm in C. Photographs and description courtesy of Elisabeth Wheeler. Line drawing by Linda Lutz-Ryan.

Not all of the petrified trees are redwoods; there are also at least five kinds of angiosperms (flowering hardwoods). Unlike conifers (softwoods), which have only tracheids serving for both support and water transport, angiosperms have two distinct cell types for these functions: vessels for water transport and fibers for support. Hardwoods thus have more complex wood anatomy. Four of the Florissant hardwoods are ring-porous, having large-diameter vessels produced early in a growing season that taper to smaller-diameter vessels formed later in the season, thus forming distinct rings. Such ring-porous woods are characteristic of highly seasonal climates with a distinct dry or cold season. The four ring-porous woods at Florissant most closely resemble *Koelreuteria* (golden-rain tree; Figure 52), *Robinia* (locust), and two types similar to *Zelkova* (Caucasian elm). These have been found only at sites outside of the main petrified forest area. It is likely that one of the woods resembling *Zelkova* corresponds to the foliage and fruits described as *Cedrelospermum*, which is an

Figure 52. These sections of wood of *Koelreuteria* (golden-rain tree) show the same planar orientations as described in Figure 51. *Koelreuteria* wood, along with most of the of angiosperms (broad-leaved hardwoods) from Florissant, is ring-porous, and this feature can be seen in the cross-section (A). Ring-porous woods have relatively large water-conducting vessels that make up most of the earlywood, and narrow vessels that make up the latewood. The vessels are the large pores (holes) visible in the cross-section (A). The radial section (B) shows the rays, made of horizontally aligned bricklike cells, with a vessel in the middle of the photograph. Two pipelike vessels are visible in the tangential section (C). Approximate dimensions of the photographs are 1.2 x 2.0 mm in A; and 0.75 x 1.0 mm in B and C. Photographs and description courtesy of Elisabeth Wheeler.

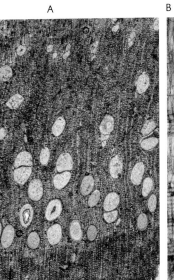
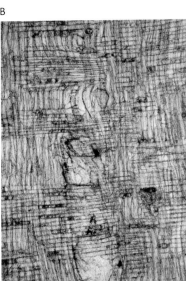
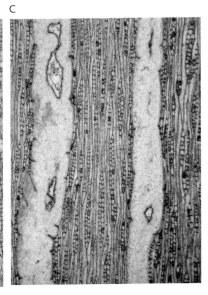

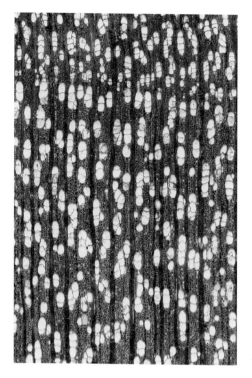

Figure 53.
Chadronoxylon was described as a new, probably extinct genus based on wood. This cross-section of the wood shows that it is diffuse-porous with scattered vessels and indistinct growth rings. It could be the wood that corresponds with the extinct leaves and fruits of *Fagopsis*, although unless all of these organs can be found in attachment, we may never know for certain. Photograph measures approximately 1.8 x 2.7 mm. Photograph courtesy of Elisabeth Wheeler.

extinct genus in the elm family and one of the two most common leaf fossils at Florissant. It is not possible to verify this relationship, however, until identifiable wood is found attached to *Cedrelospermum* leaves.

The fifth and most common of the Florissant hardwoods is placed into a new genus, *Chadronoxylon*, which is found intermixed with the *Sequoia* stumps in the main petrified forest area near the Monument headquarters. *Chadronoxylon* is diffuse-porous (i.e., the vessels are more scattered), although it does have growth rings (Figure 53). It was a large tree, up to 1.7 meters in diameter, suggesting a height of at least 30 meters. It has characteristics that occur in more than one modern plant family, and its taxonomic affinities are unclear. Could it be the wood that corresponds with the extinct genus *Fagopsis*, known from leaves that are among the most abundant at Florissant? Many characters of the wood suggest that it does not belong in the same family as *Fagopsis*—at least according to the current placement of *Fagopsis* into the family Fagaceae—but what if the unusual *Fagopsis* actually represents an extinct family? Until woody twigs with well-preserved anatomical features are found attached to leaves, we may never know whether *Chadronoxylon* and *Fagopsis* represent the same tree.

The number of plants known from wood is small compared with the number known from leaves and pollen, and undoubtedly, woods of many species were not fossilized. Whether a tree becomes petrified can be influenced by the permeability of its wood and the resistance of the wood to decay. Only trees growing along the valley bottom, in the path of the lahar mudflow, were buried. Overcollecting by early visitors may have removed unique specimens, making them unavailable for scientific study. Some combination of these factors probably accounts for the low diversity of wood types at Florissant. By contrast, fossil leaves, fruits, and pollen provide a much broader insight into the composition of the ancient forest at Florissant.

FOSSIL LEAVES, FRUITS, SEEDS, AND FLOWERS

Florissant's ancient forest is best known from the abundant impressions and compressions of leaves, fruits, seeds, cones, and flowers, all of them preserved in the shales. About 120 species can be counted from these organs alone. In some

Figure 54. Some Florissant fossils really are fluorescent! This image of a *Fagopsis* cuticle was taken using a fluorescent microscope. Cuticle is the highly resistant substance that coats all land plants, and it takes an impression of the epidermis, the outermost layer of cells on the leaf. The epidermis is composed of epidermal cells and stomata, which control the exchange of gases and water vapor in the plant. A fluorescent light was beamed onto this fossil specimen, causing the molecules in the fossil cuticle to "auto-fluoresce," emitting light in a certain frequency that appears green. The image shows the detail of the cuticle in bright green, revealing stomata, hair bases, and epidermal cells. One of the stomata is conspicuous as the bright green area with a roughly circular center. The cuticle is broken in places, and the matrix appears black. Photograph and description courtesy of Jennifer McElwain.

instances, the leaf cuticle is preserved, showing details of the outermost layer of the leaf's cells (Figure 54). Most of the leaf and fruit fossils come from trees and shrubs, and only rarely are the small herbaceous plants represented. Angiosperms dominate the flora, but conifers are also conspicuous. The examples discussed or illustrated in this chapter include only a selection of the most significant or abundant plants, and a complete list of all species is given in Appendix 1.

Plants are classified according to their reproductive strategies, with spore-producing plants such as mosses being evolutionarily primitive, and flowering angiosperms the most advanced. The more primitive plant groups at Florissant include two species of moss and one each of a horsetail and fern, although only

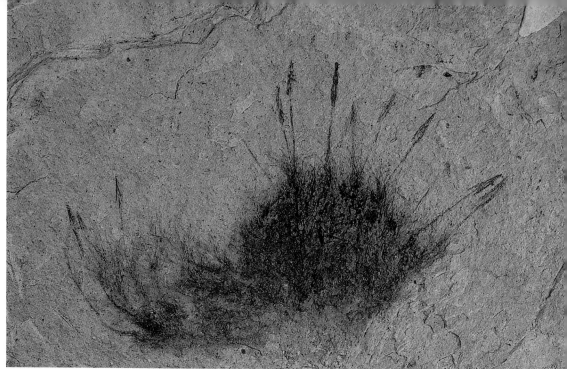

Figure 55. Mosses are uncommon in the fossil record, but this specimen of *Plagiopodopsis cockerelliae* is especially well preserved. It may have grown on a tree overhanging the lake, and then fallen into the water and become fossilized. Specimen YPM-35484 (holotype), courtesy of Peabody Museum of Natural History, Yale University.

Scale: × 4.0

Figure 56. *(right)* This jointed stem of *Equisetum florissantense* is characteristic of the horsetails, which are rare at Florissant. During the Paleozoic Era, long before Florissant, many forms of horsetails were treelike, but today, only the herbaceous *Equisetum* survives. Specimen UCM-8619 (holotype), courtesy of the University of Colorado Museum.

Scale: × 3.1

Figure 57. *Dryopteris guyottii* is the only species of fern at Florissant described from a frond. Other ferns are known as fossil spores. This fern most likely preferred the damp, shaded environments of the forest understory near the lake or along the tributary streams. Specimen WC-FL-51, courtesy of the late Paul R. Stewart, Waynesburg College.

Scale: × 0.7

the fern is common among the Florissant fossils. Mosses, because of their growth habit and lack of vascular tissue, are unlikely candidates for fossilization and are rare in the fossil record, yet the Florissant specimen of *Plagiopodopsis cockerelliae* (Figure 55) is an unusual example of a moss showing remarkable preservation. Only two specimens of horsetail are known from Florissant, belonging in the single living genus of this group, *Equisetum* (Figure 56). The horsetail would have inhabited the wet areas around the margin of the lake. *Dryopteris guyottii* is the only fern described from a frond (Figure 57), and it, too, preferred moist microhabitats.

The conifers were conspicuous members of the ancient forest, and their fossils include *Torreya, Chamaecyparis* (white cedar or false cypress), *Sequoia* (redwood), *Abies* (fir), *Picea* (spruce), and *Pinus* (pine). *Torreya* is in the family Taxaceae (yew family), and like many plants, it is known by various common names, among them "nutmeg." However, the name "nutmeg" is used for other unrelated plants as well, and *Torreya* is certainly not the source of the nutmeg used for flavoring. Other common names for *Torreya* include kaya nut, stinking cedar, and stinking yew, but other genera of conifers are also called cedars and yews. This is why it is best to emphasize the Latin generic names for plants. The fossils of *Torreya* at Florissant consist of isolated leaves, or needles, which are relatively wide and sharply pointed (Figure 58). This foliage was probably highly aromatic, as it is in modern *Torreya* trees. *Torreya* does not have woody cones, and instead the seeds develop within a fleshy plumlike cup that is not well suited for fossilization. The genus is adapted to warm temperate climates today, and is distributed in California, Florida, and southeast Asia.

Figure 58. *Torreya geometrorum* (Family Taxaceae) is a conifer that is known at Florissant only from its relatively wide, sharply pointed needles. The two narrow lines along the length of the needle are stomatal bands, where the stomata (microscopic openings that allowed the plant to breathe) were concentrated. The modern relatives of *Torreya* live in warm, moist climates. Specimen USNM-40571, courtesy of the National Museum of Natural History.

Scale: × 3.5

Figure 59. The delicate foliage of the conifer *Chamaecyparis linguafolia* (false cypress; Family Cupressaceae) consists of many small scalelike leaves that form flattened branchlets. It is a common fossil at Florissant. Judging from modern species, the Florissant *Chamaecyparis* was an aromatic tree that rivaled *Sequoia* both in stature in the tall forest canopy and in longevity. It grew in groves along with *Sequoia* in the valley bottom. Specimen UCMP-3780, courtesy of the University of California Museum of Paleontology.

Scale: natural size

One of the common conifers is *Chamaecyparis* (false cypress or white cedar), in the family Cupressaceae (cypress family). It was a tall, long-lived tree that probably grew along the valley bottom. Foliage (Figure 59) and cones are abundant, yet curiously, none of the tree stumps in the petrified forest have yet been identified as *Chamaecyparis*. The eight modern species of this genus live far from the Rocky Mountains, along the west and east coasts of North America and in eastern Asia. The fossils are similar to the living *Chamaecyparis lawsoniana* of southwestern Oregon and northeastern California, the same geographical area

Figure 60. Many branches of fossil *Sequoia* (redwood; Family Taxodiaceae) from Florissant have foliage with spreading leaves, which is also common in modern *Sequoia*, but the leaves in the fossil species usually are thinner and more delicate. This specimen of *Sequoia affinis* was once featured on a postcard in a series on "Tertiary Fossil Plants" issued by the British Museum of Natural History in the early 1920s. Specimen NHM-V.11258, courtesy of The Natural History Museum, London.

Scale: × 0.6

Figure 61. Some branches of the Florissant redwood, *Sequoia affinis*, have foliage that is flattened into the axis of the branch, with smaller, more upwardly pointed leaves. This type of foliage is seen less commonly in branches of the modern coast redwood, *Sequoia sempervirens*, but is more characteristic of the Sierra redwood or giant sequoia, which is placed into another genus, *Sequoiadendron giganteum*. Could it be that the Florissant *Sequoia affinis* is ancestral to both of the modern redwoods? Specimen WC-FL-36, courtesy of the late Paul R. Stewart, Waynesburg College.

Scale: × 0.5

where modern *Sequoia* grows. This modern region represents a refugium where the relatives of several plants, more widely distributed during the Tertiary Period, still survive as relicts.

Sequoia (redwood), common as stumps in the petrified forest, is found also as foliage, cones, and pollen at Florissant. Like the wood, these show distinct

Figure 62. Female cones of *Sequoia affinis* are only 50–70 percent of the length of modern redwood cones. Along with other differences in the wood, foliage, and pollen, this shows that the Florissant redwood is distinct from its modern relative, *Sequoia sempervirens*, the coast redwood of California. Specimen UCM-34188, courtesy of the University of Colorado Museum.

Scale: × 1.2

differences from the modern coast redwood of California. The foliage of the fossils is thinner and more frequently flattened into the axis of the branch, or appressed, than the modern species (Figures 60 and 61), and the female cones (Figure 62) are only 50–70 percent of the size of modern *Sequoia* cones. Although the Florissant species (*Sequoia affinis*) is assigned to the same genus as the modern coast redwood (*Sequoia sempervirens*), these features can be used to distinguish the two. *Sequoia affinis* may have been ancestral to *Sequoia sempervirens,* but it also may have been the ancestor of the modern *Sequoiadendron giganteum* (Sierra redwood or giant sequoia), which is classified as a different genus that is not well documented in the fossil record until the Miocene Epoch about 16 million years ago. Both of these modern genera are

Figure 63. Among Florissant's leaf and seed fossils, this winged seed of *Abies rigida* (Family Pinaceae) is the only known specimen of the fir, although *Abies* is more abundant in the fossil pollen record at Florissant. This tall tree with wind-dispersed seeds would seem a likely candidate for deposition in the fossil record. Its rarity at Florissant is probably because it was uncommon or absent in the immediate vicinity around the lake, growing instead at higher elevations. It is not always easy to correlate abundance in the fossil record with abundance in the ancient forest. Specimen UCMP-3772, courtesy of the University of California Museum of Paleontology.

Scale: × 5.8

endemic to the western United States, with *Sequoia* restricted to coastal California and southern Oregon, and *Sequoiadendron* living only in the Sierra Nevada mountains of California. Of the two, *Sequoia* clearly shows the closest similarity to the fossil species from Florissant. *Sequoia* is traditionally placed into the family Taxodiaceae (bald cypress family), although more recent work suggests a merger of that family into the family Cupressaceae (cypress family), along with *Chamaecyparis*. Like the *Chamaecyparis* at Florissant, *Sequoia* was a large, long-lived, dominant conifer that grew along the moist valley bottom.

The Pinaceae (pine family) is well represented as foliage and seeds of *Pinus* (pine), *Picea* (spruce), and *Abies* (fir). Pine is abundant as needles, seeds, and cones, but spruce and fir are found only as rare seeds. In fact, there is only one known seed specimen of the fir (Figure 63), and it is a type that resembles the modern *Abies bracteata* (bristlecone fir), which is restricted to a small area near the central California coast today. It is also the same type of fir that occurs in the Oligocene Creede flora. Spruce is more common at Florissant, but it, too, is known only from seeds. Spruce and fir probably grew in the cooler volcanic highlands around the Florissant basin, or in shaded ravines around the lake valley, and their seeds were carried into the ancient lake by tributary streams from the surrounding uplands. By contrast, their pollen was widely dispersed by the wind, and for that reason, it is more abundant in the fossil record at Florissant.

As many as five or six species of *Pinus* occur at Florissant, but the actual number is difficult to determine because of the different isolated organs of foliage, seeds, cones, and cone scales. Indeed, the various names that have been given to the Florissant pines serve as an excellent example to show why paleobotanists need to avoid giving the same species name to different, unattached fossil organs of the same genus. For example, MacGinitie assigned a cone, seeds, and foliage

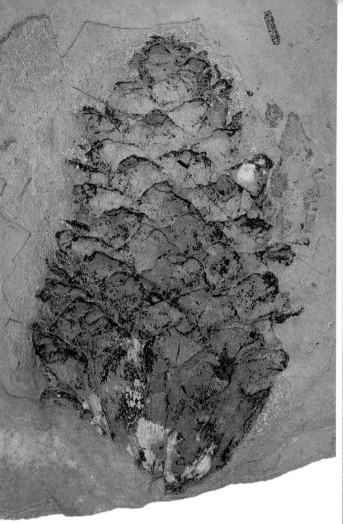

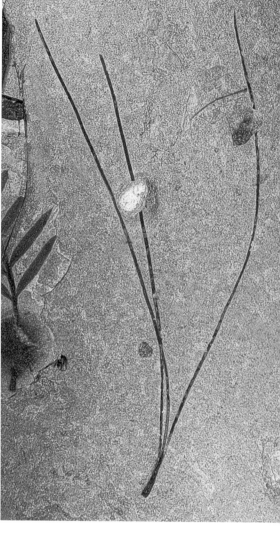

Figure 64. The tips of the cone scales of this female cone of *Pinus florissantii* (Family Pinaceae) are characteristic of the subgenus Haploxylon (soft pines). The foliage of soft pine (*Pinus wheeleri*, based on foliage only) also is known from Florissant, but it is uncertain whether it belongs to the same species as this cone. *Pinus florissantii* is related to the modern *Pinus flexilis*, the limber pine that grows in the Florissant area today. The specimen illustrated here is among the many superb fossils that were collected by the Princeton Scientific Expedition of 1877. Specimen USNM-387525 (holotype), courtesy of the National Museum of Natural History.

Scale: natural size.

Figure 65. *Pinus macginitiei* is distinguished by the three-needled fascicle (bundle of needles) with a bulbous base enclosed within a persistent sheath (the darkened area at the bottom), indicating that the species belongs to the subgenus Diploxylon (hard pines). One specimen of a hard pine cone (formerly *Pinus wheeleri*, cone only) and various specimens of winged seeds are also known from Florissant, but it is uncertain whether these belong to the same species as this foliage. UCMP-3776 (holotype), courtesy of the University of California Museum of Paleontology.

Scale: × 0.9

Figure 66. Pine seeds are usually winged, and those with a detachable, articulated seed body such as this are typical of the hard pines (Subgenus Diploxylon). The seed body had already become detached in this specimen, leaving only the two hooks at the proximal end of the wing. This type of seed has been referred to *Pinus macginitiei*, but it is difficult to know for certain whether these seeds and the foliage illustrated in Figure 65 actually belong to the same species, because they have not been found in attachment. Specimen UCMP-141988, courtesy of the University of California Museum of Paleontology.

Scale: × 2.0

all to *Pinus florissantii,* a species originally described by Lesquereux from a cone (Figure 64). We now know that this cone comes from the group of pines known as soft pines (subgenus Haploxylon), and that the seeds and foliage come from a hard pine (subgenus Diploxylon). The name for *Pinus florissantii* must always apply first to the cone, because that is what Lesquereux designated as the holotype (name-defining) specimen in his original 1883 description of the species. The foliage and seeds (Figures 65 and 66) were renamed as *Pinus macginitiei,* a two- to three-needled hard pine, but with no proof that even they belong in the same species. A similar example is *Pinus wheeleri,* originally described by Cockerell from two specimens, one a cone and the other foliage. The cone is now known to be a hard pine, and the foliage a soft pine. Unless attachments can be proven, it is difficult to place different plant organs into the same species with confidence. The various species of pine probably occupied a range of habitats around the Florissant basin. Most pines probably lived in moderate abundance on the dry slopes surrounding the lake, but some may have lived closer to the lake whereas others were more distant and grew at higher elevations.

Although the conifers were important at Florissant, it is the broad-leaved flowering plants that dominated and were most diverse in the ancient forest. The laurel family (Family Lauraceae) is represented by two genera at Florissant, based on leaves (Figure 67) that resemble those of *Lindera* (or *Litsea*) and *Persea,*

Figure 67. This leaf resembles leaves typical of the laurel family (Family Lauraceae), such as *Lindera* and *Litsea*. Compared with leaves of many other families, laurels are often more difficult to identify accurately to genus. Laurels are common in tropical climates but also range into warm temperate regions. Like many leaves that are adapted to warmer climates, they typically have entire (smooth) margins, which provide important evidence for reconstructing past temperatures. Specimen UCMP-3629 (holotype), courtesy of the University of California Museum of Paleontology.

Scale: × 0.75

Figure 68. This leaflet of *Mahonia marginata* (Family Berberidaceae) is similar to the modern Oregon grape. Like its modern relatives, the plant was a shrub, and the evergreen leaves had stout, asymmetrical leaflets with spinose teeth, perhaps as an adaptation to defend against herbivory. Specimen USNM-330770, courtesy of the National Museum of Natural History.

Scale: × 1.5

Figure 69. The male flowering head of *Fagopsis longifolia* (representing an extinct genus in the Fagaceae, or beech family) consists of many stamens, some of which contain fossilized pollen. Specimen YPM-30249, courtesy of Peabody Museum of Natural History, Yale University.

Scale: × 1.2

although these identifications need closer scrutiny to verify their validity. *Lindera* is similar to the modern wild allspice or spicebush, a shrub in the forests of the central and eastern United States. *Persea* includes modern plants such as the red bay and avocado. It is an evergreen shrub or small tree that is largely tropical today but includes species that range north into warm temperate areas along the Atlantic coast of the United States. Both of these genera are rare as fossils at Florissant, perhaps in part because they were evergreen and did not shed their leaves as frequently as the more common deciduous plants.

The family Berberidaceae (barberry family) is represented by three species

Figure 70. The attachment of different organs of an extinct plant provides important evidence for piecing together the whole plant. This specimen shows a female flowering head attached to a twig bearing the foliage of the extinct genus *Fagopsis*. Following pollination, such flowering heads matured to form fruiting heads (Figure 72). The leaves have straight secondary veins and large prominent teeth. *Fagopsis* is one of the two most abundant plants at Florissant. Specimen YPM-30121, courtesy of Peabody Museum of Natural History, Yale University.

Scale: × 1.1

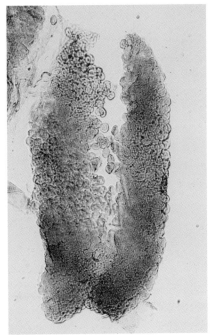

Figure 71. Microscopic examination of a sample prepared from a male flowering head of *Fagopsis longifolia* reveals a pollen sac containing hundreds of tiny pollen grains. The pollen provides yet another example to show how an extinct plant can be reconstructed when multiple organs are found in attachment. Total length of pollen sac is approximately 0.7 mm. Photograph courtesy of Steven Manchester.

Scale: × 100

of *Mahonia* (Oregon or holly grape) at Florissant. *Mahonia marginata* (Figure 68) is similar to the modern Oregon grape, representing a subgenus of *Mahonia* that today lives in North America only west of the Cascade Mountains and is the only subgenus of *Mahonia* in Asia. The other two species of Mahonia (*M. obliqua* and *M. subdenticulata*) are more closely related to modern species that live in the dry southwestern United States. All of these species were probably shrubs that grew in the forest understory, or on the drier slopes around the lake basin. They are primarily temperate in modern distribution and range into cold climates.

The family Fagaceae (beech family) includes *Quercus* (oak) and the extinct genus *Fagopsis*. Some leaves have been referred to *Castanea* (chestnut), although

Figure 72. The fruiting head of *Fagopsis longifolia* developed as the female flowers (Figure 70) matured. About 40 fruit wedges (Figure 73) encircled the peduncle (the axis of the fruiting head) in what was probably a spiraling arrangement, and in this example, the upper ones have been shed. Protruding bracts, and scars on the surface of the peduncle, show where detachment of the fruit wedges occurred. Specimen USNM-334283, courtesy of the National Museum of Natural History.

Scale: × 2.3

Figure 73. *(above) Fagopsis longifolia* fruit wedges cling together to form a circular or C-shaped string that unravels as it is shed from the fruiting head (Figure 72). Each individual fruit wedge is a cupule, similar to the cup-like structure of an acorn that holds the nut. Microscopic examination shows that three tiny fruits, or nuts, occur near the base of each fruit wedge. Isolated fruit wedges have not been found as fossils, and apparently it was the circular string of clinging wedges that formed the entity for dispersal. Dispersal of the strings probably occurred by wind. Specimen USNM-334278, courtesy of the National Museum of Natural History.

Scale: × 2.2

Figure 74. *(below)* The reconstruction of *Fagopsis longifolia* (Family Fagaceae) is possible only because different organs of the plant are found attached to one another in the fossil record at Florissant. This illustration shows a female flower head (upper left); mature fruiting head before shedding of the fruit wedges (upper right); fruiting head during shedding of the fruit wedges (lower right); pollen-producing male flower head with stamens (lower left); and young leaf (middle left). From Manchester and Crane (1983), reproduced with permission from the American Journal of Botany.

this identification has been questioned and the leaves may instead represent another species of *Quercus*. The most common member of the Fagaceae—and, along with *Cedrelospermum*, one of the two most common plants at Florissant— is the extinct genus *Fagopsis*, represented by the single species *Fagopsis longifolia*. The genus is known from only three other fossil floras, all from North America and of Eocene age. *Fagopsis* fossils are significant at Florissant because some of them show the physical attachment of the different plant organs (such as foliage, flowers, fruits, and pollen), making it possible to reconstruct an extinct genus from multiple organs representing various stages of the plant's reproductive cycle. The wood, which has not been found in attachment with the other organs, remains a mystery, although the wood referred to *Chadronoxylon* may be a good candidate.

Similar to modern genera in the Fagaceae, *Fagopsis* had separate male and female flower clusters, known as staminate and pistillate inflorescences, respectively (Figures 69 and 70). Samples that have been taken and prepared from the staminate inflorescences show that pollen is preserved in some specimens (Figure 71), clearly showing which type of pollen corresponds with the extinct plant. The pistillate inflorescences matured into infructescences (fruit-bearing structures; Figure72) bearing strings of fruit wedges that unraveled as they were shed (Figure73). Each fruit wedge is homologous to the cuplike structure, or cupule, that contains the nut in an acorn of an oak, and each individual fruit wedge in *Fagopsis* contains three tiny nuts. The strings of clinging fruit wedges evidently were adapted for wind dispersal, unlike the more typical dispersal by animals for modern members of the family, such as *Quercus* (oak) and *Fagus* (beech). These features of the infructescence, which distinguish *Fagopsis* from its modern relatives, may have evolved in response to selective pressures favoring wind dispersal. Although *Fagopsis* possesses combined characteristics of leaves, inflorescences, pollen, and infructescences that show that it belongs in the family Fagaceae, its unusual features, such as the tiny nuts and unique fruit wedges, clearly distinguish it from modern genera in the family. The rare fossils that preserve attached organs of *Fagopsis* make it possible to reconstruct this extinct genus (Figure 74).

With nine species, *Quercus* is the most diverse plant genus at Florissant, and it is known from both leaves and acorns (Figures 75 and 76). Fossils of oak are not common, probably because the trees preferred the drier, well-drained slopes surrounding the valley, more distant from the lake itself. Many modern oak species show tremendous variation in leaf form, and different species often hybridize with one another. For these reasons, it is particularly difficult to differentiate species from the fossil record, and the actual number of oak species

Figure 75. There are several species of oaks at Florissant, including this *Quercus predayana* (Family Fagaceae). The darkened, carbonized areas are indicative of a leaf with a thick texture, suggesting that this was an evergreen oak, probably similar to modern species of "live oak" in California, Arizona, and northwest Mexico. Oak leaves are uncommon at Florissant, probably because the trees preferred the drier, well-drained slopes more distant from the lake. Specimen USNM-40551 (syntype), courtesy of the National Museum of Natural History.

Scale: × 1.9

Figure 76. This fruit of *Quercus* sp. (oak) shows the nut of the acorn (blackened area) attached to the scaly cup (lighter brown half of the specimen). Specimen WC-FL-6, courtesy of the late Paul R. Stewart, Waynesburg College.

Scale: × 2.1

based on fossil leaves is somewhat arbitrary. Many of the Florissant oaks are similar to modern species that form the dry oak forests of northern Mexico and the southwestern United States. Modern oaks include both evergreen and deciduous species. This can be difficult to determine from the fossil record, other than by the clues of leaf texture or nearest living relatives, but it is likely that the Florissant oaks represent a mixture of both evergreen and deciduous species.

The family Betulaceae (birch family) is known at Florissant not from its modern genera such as birches and alders, but from two extinct fossil genera—*Paracarpinus,* based on leaves, and *Asterocarpinus,* based on fruits. In fact, these two fossil genera probably belonged to a single biological genus, but because

FOSSIL PLANTS 95

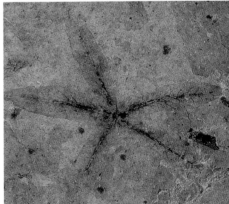

Figure 77. *Paracarpinus fratorna (top)* is an extinct leaf genus belonging to the Betulaceae (birch family). These leaves are almost indistinguishable from those of modern *Carpinus* (hornbeam), but *Paracarpinus* is considered to be the leaf corresponding to fruits of the extinct genus *Asterocarpinus* (Figure 78). The leaves are characterized by the straight, closely spaced secondary veins and long, sharply pointed teeth. This leaf also shows extensive insect feeding damage (Figure 42). Specimen UCMP-198423, courtesy of the University of California Museum of Paleontology.

Scale: × 1.3

Figure 78. *(below) Asterocarpinus* is an extinct genus in the Betulaceae (birch family) described from fruits that typically have five wings radiating from a central nutlet. The co-occurrence at Florissant and elsewhere of *Asterocarpinus* fruits with *Paracarpinus* leaves (Figure 77) suggests that these two fossil genera actually belonged to a single biological species. This relationship is difficult to prove unless the leaves and fruits are found attached to one another in the fossil record, so they are given separate names as fossils. *Paracarpinus* leaves are indistinguishable from modern *Carpinus* (hornbeam), but *Asterocarpinus* fruits are unique. This is a good example to show differential rates of evolution between different plant organs, with the leaves little changed but the fruits showing more rapid evolution. This specimen is also a good example to show how the name of a particular fossil can change as the specimen is reexamined and published by different authors. It was first described by Cockerell in 1908 as *Buettneria? perplexans* (Family Sterculiaceae), later assigned by MacGinitie in 1953 to *Petrea perplexans* (Family Verbenaceae), and most recently referred by Manchester and Crane in 1987 to *Asterocarpinus perplexans* (Family Betulaceae), which represents an extinct new genus. Specimen UCMP-3622, courtesy of the University of California Museum of Paleontology.

Scale: × 2.3

they are not found attached to one another, they are described separately on the basis of these isolated organs. As we have already seen, this somewhat confusing situation is often a problem in paleobotany when the leaves and fruits are not found attached. *Paracarpinus* (Figure 77) leaves are almost identical to those of modern *Carpinus* (hornbeam), but the fruits of *Carpinus* are unknown from either Florissant or other sites where *Paracarpinus* occurs. By contrast, the distinctive fruits of *Asterocarpinus* (Figure 78), consisting of four to seven wings radiating from a central nutlet, do occur at the same sites with *Paracarpinus*. Both *Paracarpinus* and *Asterocarpinus* share characteristics of the subfamily Coryleae. These similar associations and characters strongly support the suggestion that the two fossil genera represent what was actually the same biological genus. *Asterocarpinus-Paracarpinus* is a good example of differential rates of evolution between different organs, with the leaves similar to modern *Carpinus* but with the fruits showing unique characteristics of a new genus. Fossil evidence suggests that *Asterocarpinus-Paracarpinus* was a plant that only existed for about 2–3 million years, during the latest Eocene and early

Figure 79. This compound leaf of the hickory *Carya libbeyi* (Family Juglandaceae) has nine attached leaflets and is one of the largest complete fossils ever collected at Florissant. Florissant is the earliest known fossil record for hickories, which later spread into Europe and Asia. Plants such as this provide important evidence for biotic exchanges across ancient land connections of North American with Europe and Asia. Specimen UCMP-3601, courtesy of the University of California Museum of Paleontology.

Scale: × 0.5

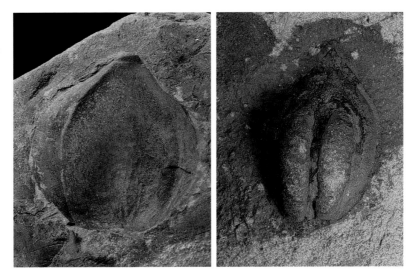

Figure 80. These two specimens show the outside surface (left) and inside nut (right) of the hickory fruit *Carya florissantensis*. Hickory fruits have smooth outer surfaces, and the nut is similar to their modern representative, the pecan. This species of fruit is thought to correspond to the leaf of *Carya libbeyi*, but they have different names because the two organs have not been found attached. Specimen on the left is USNM-40558, courtesy of the National Museum of Natural History. Specimen on the right is UCMP-3603 (paratype), courtesy of the University of California Museum of Paleontology.

Scale: ×1.6

Oligocene. Like many modern genera in the birch family, *Asterocarpinus-Paracarpinus* probably grew in moist environments near the margin of the lake. The attractive flowerlike fruits were well adapted for wind dispersal.

The family Juglandaceae (walnut family) is represented by *Carya* (hickory) and *Juglans* (walnut). The walnut is known only from nutshell impressions, whereas the hickory is known from leaves, nuts, and male flowers containing pollen. The hickory leaves are described as *Carya libbeyi* and the fruits as *Carya florissantensis* (Figures 79 and 80). These leaves and nuts may well belong to the same species, but they were named separately because the two organs have not been found attached. The Florissant hickory is the earliest known fossil record for the genus, which evidently first appeared in North America and later spread into Europe and eventually Asia. The crossing into Europe by the Oligocene Epoch supports evidence for biotic exchange over a land connection across the North Atlantic during this time. The hickories became very diverse in the fossil record of Europe, but today they are native only to eastern North America, northeast Mexico, and eastern Asia.

Like their modern relatives, the poplars and willows (Figures 81 and 82), in the family Salicaceae (willow family), probably lived exclusively in the wet areas along streams and near the lake. This close proximity to the lake explains why the single species of *Populus* (poplar) is so common as fossils at Florissant. In contrast, *Salix* (willow) is rare in the fossil record in spite of the fact that four species are recognized. Perhaps the poplars were better able to compete for the limited lakeside habitat.

Florissantia is an unusual extinct genus probably belonging in the family

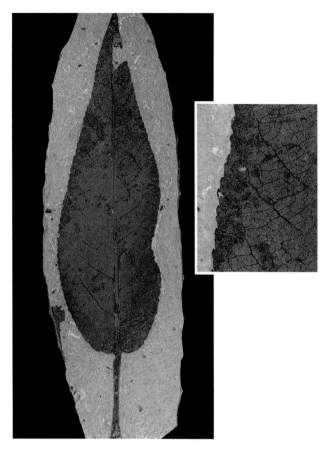
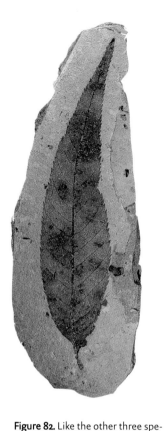

Figure 81. The poplar *Populus crassa* (Family Salicaceae) is one of the largest and most common leaves found at Florissant. The trees required ample moisture and grew in wet areas along streams and around the margin of the lake, where they were good candidates for fossilization. The leaves can be distinguished by a pair of secondary veins arising at the base, venation with net-shaped higher-order veins (see inset), and small glands on the teeth. Specimen USNM-40549, courtesy of the National Museum of Natural History

Scale: × 0.6

× 1.7 (inset)

Figure 82. Like the other three species of Florissant willows, fossils of *Salix ramaleyi* (Family Salicaceae) are surprisingly uncommon for a plant that would be expected to have lived close to lakes and streams. The slightly wavy course of the secondary veins is one feature that often characterizes willow leaves. Specimen USNM-40557, courtesy of the National Museum of Natural History.

Scale: × 0.8

Sterculiaceae. It is known from star-shaped flowers or fruits from Florissant and elsewhere. Like so many of the Florissant fossil plants, it has been identified under different names over the years, originally as *Porana* (Family Convolvulaceae) by Lesquereux and later as *Holmskioldia* (Family Verbenaceae) by MacGinitie, both names of modern genera. Closer examination of the specimens reveals characteristics of the five-lobed flower that show that the plant

Figure 83. Flowers are relatively rare in the fossil record because of their delicacy, but those of *Florissantia speirii* retained the toughened calyx as they matured into fruits. The star-shaped matured calyx formed a tiny umbrella-like structure that enhanced wind dispersal and transported these organs into the lake. Although all of the Florissant plants represent extinct species, *Florissantia* is one of several extinct genera. Its corresponding leaf remains a mystery. Specimen UCMP-3619, courtesy of the University of California Museum of Paleontology.

Scale: × 3.4

belongs in the extinct genus *Florissantia*, having affinities with the family Sterculiaceae. Most of the Florissant specimens are fruits that developed from the flower's ovary and persistent calyx as it matured (Figure 83). The flowers have features suggesting pollination by insects or birds, and the fruits were probably wind dispersed. The corresponding leaf is unknown. The plant is known from the Eocene and Oligocene of western North America but does not appear in the fossil record again until the Miocene of eastern Asia. These occurrences on different continents—separated in time by millions of years—raise intriguing questions about the dispersal of *Florissantia*. The available fossil evidence lacks the resolution to provide clear answers to these questions, although it is easy to hypothesize that *Florissantia*, like other plants known from Florissant, reached Asia from North America by crossing Beringia, the ancient land connection between Alaska and Siberia.

The family Ulmaceae (elm family) is known from both leaves and fruits of *Ulmus* (elm) and the extinct genus *Cedrelospermum*, and from leaves of *Celtis* (hackberry). The Florissant fruits now assigned to *Cedrelospermum* were once considered to belong to *Banksites* or *Lomatia*, in the Southern Hemisphere family Proteaceae. Closer examination of the fruits (Figure 84) indicates that they

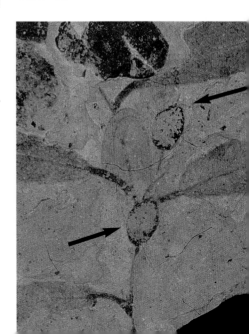

Figure 84. This fruit of *Cedrelospermum lineatum,* an extinct relative of the elms (Family Ulmaceae), was once thought to be the seed of *Lomatia* (Family Proteaceae), a plant that grows today only in South America and Australia. That misidentification was cited as evidence that the Florissant plant community had some affinity with modern plants in the Southern Hemisphere. Closer examination later revealed that the stigmatic area at the tip of the wing—evident in this photograph as the darkened area where the veins converge—is characteristic of the Ulmaceae. This, along with other specimens found in attachment with foliage (Figure 85), shows that what were first thought to be *Lomatia* seeds are in fact fruits belonging in the elm family. Specimen USNM-387567 (lectotype), courtesy of the National Museum of Natural History.

Scale: × 4.3

Figure 85. Representing an extinct genus in the Ulmaceae (elm family), *Cedrelospermum lineatum* is one of the two most common plant fossils at Florissant. The leaves of this plant were formerly identified as *Zelkova,* a modern genus in the elm family. When specimens were found showing this type of leaf attached to the distinctive fruits of the extinct genus *Cedrelospermum* (Figures 84 and 86), the name of the leaves was changed to match. The leaves are often asymmetrical and, like other members of this family, have veins that characteristically branch from the bottom sides of secondary veins and end in the sinuses between teeth (see inset). Specimen UF-7279, courtesy of the Florida Museum of Natural History, University of Florida.

Scale: × 2.1
  × 4.7 (inset)

Figure 86. Fossils that preserve the attachment of leaves and fruits, such as this specimen of *Cedrelospermum lineatum,* are important for proving that these two organs belonged to the same plant species. This example shows a branch with ulmaceous leaves (formerly considered to be *Zelkova,* or Caucasian elm) attached to the fruits (indicated by the arrows) of the extinct genus *Cedrelospermum.* Specimen YPM-23956A, courtesy of Peabody Museum of Natural History, Yale University.

Scale: × 3.2

actually have characteristics of the elm family, but represent an extinct genus. Leaves (Figure 85) described as *Zelkova* (a modern genus in the elm family) were discovered in attachment to the fruits of *Cedrelospermum* (Figure 86), proving that these leaves also belonged to the same extinct genus. The leaves and fruits are now known as *Cedrelospermum lineatum*. *Cedrelospermum* serves as a good example of multiple-organ reconstruction showing differential evolution of plant organs, with conservative evolution of leaves (*Cedrelospermum* leaves are similar to those of modern *Zelkova*), but more rapid evolution of fruits (*Cedrelospermum* fruits are unique for the elm family). *Cedrelospermum* is one of the two most common plant species at Florissant. It may have been an abundant tree around the margin of the lake, and the small-seeded fruits with wings adapted for wind dispersal suggest that it was an early successional plant that could have readily colonized the disturbed volcanic terrain. The genus is known from other fossil sites, including Eocene and Oligocene floras in western North America, as well from Eocene to Miocene floras in Europe. As with *Carya,* the shared distribution of this extinct genus between North America and Europe supports the evidence for a north Atlantic land connection between these two continents during the Eocene.

Morus (mulberry), in the family Moraceae, is known only from leaves (Figure 87). The clusters of fleshy fruits with tiny seeds were unlikely candidates for fossilization at Florissant and have not been found, but undoubtedly they would have attracted frequent visits by birds and other small animals. Mulberries are deciduous trees, and although many of the modern species grow in tropical areas, others live in temperate climates.

Figure 87. Leaves of the fossil mulberry, *Morus symmetrica* (Family Moraceae), have various shapes just as they do in modern mulberries. It is unusual to find two unattached leaves of the same species so close to one another, and perhaps these leaves already had clung together in the forest litter before they were transported into the lake. Specimen UCMP-3808, courtesy of the University of California Museum of Paleontology.

Scale: × 1.1

Figure 88. *Humulus florissantella* (hops), in the Cannabaceae (marijuana family), is preserved as leaves that have a somewhat square-shaped notch in the base and U-shaped notches between the lobes. The Florissant leaves were originally described as *Vitis* (grape)—which does also occur in the fossil record at Florissant—but closer examination reveals characteristics of *Humulus*. The genus is still represented in the modern forest at Florissant. Specimen USNM-33723, courtesy of the National Museum of Natural History.

Scale: × 1.1

Humulus (hops), in the family Cannabaceae (marijuana family), is known at Florissant from its leaves (Figure 88). These leaves were originally identified as *Vitis* (grape), but characteristics such as the shape of the lobes and teeth, and the U-shaped notches between lobes, indicate that the leaves actually are more similar to those of hops. At least one of the fossil leaves originally identified as a grape, however, does appear to be validly identified. *Humulus* was a climbing vine and its fruits are used today for flavoring beer.

Ribes (currant, or gooseberry) is the only genus representing the family Grossulariaceae at Florissant. The plant would have been a low shrub and may have had thorny branches. It is one of a few genera from Florissant's fossil record that can still be found growing at Florissant today.

The family Rosaceae (rose family), with 13 species in 9 genera, is the most diverse family at Florissant and includes *Amelanchier* (serviceberry; Figure 89), *Cercocarpus* (mountain mahogany; Figure 90), *Crataegus* (hawthorn; Figure 91), *Holodiscus* (ocean spray), *Malus* (apple; Figure 92), *Prunus* (plum or cherry), *Rosa* (rose; Figures 93 and 94), *Rubus* (blackberry or raspberry), and *Vauquelinia* (rosewood). Most of these were shrubs and smaller trees that grew in the understory of the forest or on dry slopes. All of the Florissant Rosaceae are placed into modern genera, although differences in the fruit suggest that the fossils referred to *Cercocarpus* may actually represent an extinct genus. The rose family had already diversified in the uplands of the middle Eocene 10 to 15 million years earlier than Florissant. Most of these genera of Rosaceae are temperate plants still common in the modern flora of the Rocky Mountains.

Another family well represented at Florissant is the Fabaceae (legume family, also known as Leguminosae; Figures 95–98). This is a large family today, and many different genera have leaves that look similar and are difficult to distinguish. Identification of legume leaves from the fossil record is therefore more challenging than it is for some other families, as is reflected in the names that are used for some legume genera at Florissant. For example, *Caesalpinites*,

 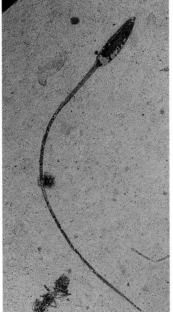 

Figure 89. The serviceberry, *Amelanchier scudderi*, is a member of the rose family. The leaves are typically serrated along the apex, but not along the base. The serviceberry was a shrub or small tree, and its fruits were probably a food source for small animals. The genus is still a common plant in the modern flora of the Rocky Mountains. Specimen UCMP-3867, courtesy of the University of California Museum of Paleontology.

Scale: × 1.2

Figure 90. *Cercocarpus myricaefolius* (mountain mahogany) is a dry-adapted member of the rose family known from distinctive fruits such as this, as well as from leaves. The peculiar fruit consists of a nutlet enclosing a single seed, and a long, tail-like style. The styles in modern *Cercocarpus* fruits are feathery and hairy, but the fossils lack these features, probably because the delicate hairs clung to the axis of the style during immersion in the lake water. Specimen UCMP-3815, courtesy of the University of California Museum of Paleontology.

Scale: × 2.1

Figure 91. With three species, *Crataegus* (hawthorn) is Florissant's most diverse genus in the rose family. The margin of *Crataegus copeana* leaves varies from being serrated to being deeply lobed, and this specimen is intermediate in showing both serrations and shallow lobes. The hawthorns were shrubs or small trees, and they may have been among the opportunistic plants that first colonized disturbed sites. Their branches possessed long spiny thorns which are occasionally found as fossils. Specimen UCMP-3610, courtesy of the University of California Museum of Paleontology.

Scale: × 0.85

Figure 92. Leaves identified as *Malus florissantensis* (Family Rosaceae) indicate that wild apples grew in the Florissant forest, providing an important source of food for birds and mammals, and a home for the developing larvae of certain insects. Specimen UCMP-3751, courtesy of the University of California Museum of Paleontology.

Scale: × 0.75

Figure 93. *Rosa hilliae* (rose), known from compound leaves, was a thorny shrub that grew in the forest understory. This and other genera of the rose family diversified in the relatively cooler upland regions of the Eocene, and by the time of the global climatic deterioration near the Eocene-Oligocene boundary, they were already well-adapted to survive the more widespread cool conditions that have prevailed since. Indeed, roses still live in the frigid environment at Florissant today. Specimen UCMP-3855, courtesy of the University of California Museum of Paleontology.

Scale: × 1.3

Figure 94. Some members of the family Rosaceae (rose family) have prickly thorns on their stems, which are commonly found as fossils at Florissant. This is probably the thorn of a rose (*Rosa*). Specimen UCMP-3716, courtesy of the University of California Museum of Paleontology.

Scale: × 3.4

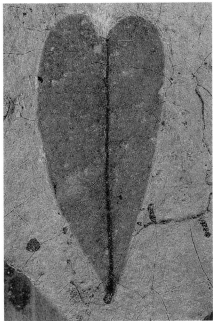

Figure 95. This small compound legume leaf (Family Fabaceae) was referred by MacGinitie to *Prosopis linearifolia*, although closer examination suggests that it probably belongs to another genus in this family. Many of the fossil plants and insects from Florissant are in need of further research in order to make more accurate identifications. Specimen YPM-11597, courtesy of Peabody Museum of Natural History, Yale University.

Scale: × 1.4

Figure 96. The legumes (Family Fabaceae) can be difficult to identify accurately, and this leaflet of *Caesalpinites coloradicus* has features in common with more than one modern genus. The uncertainty in the identification is resolved by placing it into the form genus *Caesalpinites,* which is a generic name for fossils that share characteristics with more than one modern genus. The notched apex may have formed during a brief drought that interrupted the leaf's normal development. Specimen USNM-387571 (syntype), courtesy of the National Museum of Natural History.

Scale: × 2.6

Figure 97. *(below) Cercis parvifolia* (redbud) is a member of the family Fabaceae (legumes). The small, smooth-margined leaves have prominent lateral veins arising from the base. Specimen USNM-315289, courtesy of the National Museum of Natural History.

Scale: × 3.4

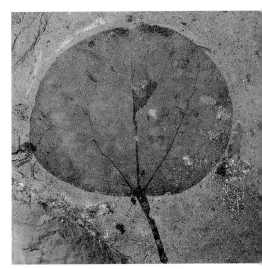

Figure 98. This beanlike legume pod (Family Fabaceae) has been referred to *Prosopis linearifolia,* but it needs more careful examination before it can be identified accurately. It may correspond to one of the legume leaves from Florissant, but it is impossible to know for certain which one unless leaves and fruits can be found in attachment. Specimen UCM-8623, courtesy of the University of Colorado Museum.

Scale: × 1.3

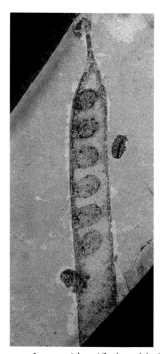

Leguminosites, and *Phaseolites* are "form genera," which are names given to fossil material as a means of showing the uncertainty in accurately placing the fossils into any one living genus. This does not necessarily mean that these are extinct genera, only that the fossils assigned to these names have affinities with more than one modern genus. Other genera of Fabaceae, such as *Cercis* (redbud; Figure 97) and *Robinia* (locust), can be assigned more confidently to living genera. Fossil fruits of the beanlike pods characteristic of the family are also found (Figure 98), but these too can present problems for accurate identification. Several of the Florissant legumes appear similar to forms that inhabit dry subtropical areas today.

Leaves identified as bladdernut, *Staphylea* (Family Staphyleaceae), are common at Florissant (Figures 99 and 100). The leaves are compound, with three leaflets. Modern *Staphylea* has bladderlike fruits that are wind dispersed, making them likely candidates for fossilization, but the absence of such fruits at Florissant casts some doubt on the validity of the identification of these fossil leaves as *Staphylea.* Fossil plants are always more confidently identified when more than one organ can be found in the fossil record. Modern bladdernuts are shrubs or small trees that grow in the moist forest understory in temperate areas throughout the Northern Hemisphere.

The family Sapindaceae is diverse at Florissant and includes fossils assigned to the genera *Cardiospermum* (balloon vine), *Dodonaea* (hop bush), *Koelreuteria* (golden-rain tree), *Sapindus* (soapberry), and *Thouinia* (guasimilla roja). For the most part, the members of this family live in subtropical to tropical climates today, and many have compound leaves. The deeply lobed leaflets of *Cardiospermum* show a distinct type of specialized hole feeding (Figure 46) that was caused by a specific insect pest, although which of Florissant's many insects was responsible for making these holes is uncertain. Modern *Cardiospermum* is a climbing vine that lives in subtropical climates. *Koelreuteria* is a medium-sized tree that is presently restricted to eastern Asia. It does well in temperate climates and is common at Florissant both as fruits (Figure 101) and compound leaves (Figure 102). *Sapindus* is represented by leaf fossils similar to the soapberries that

Figure 99. Leaves identified as the bladdernut, *Staphylea acuminata* (Family Staphyleaceae), are compound with three leaflets each. The small size of these apparently immature leaves suggests that this branchlet may have blown off the tree during a windstorm, before the leaves had fully matured. Specimen UCM-18541, courtesy of the University of Colorado Museum.

Scale: natural size

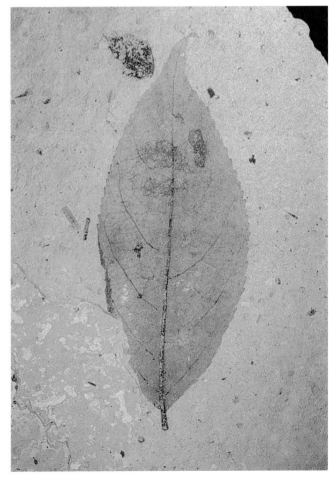

Figure 100. An isolated leaflet of the bladdernut, *Staphylea acuminata* (Family Staphyleaceae), shows small teeth and secondary veins that arc upward as they approach the margin. *Staphylea* is one of the more common leaves found at Florissant. Specimen UCM-5176, courtesy of the University of Colorado Museum.

Scale: × 1.1

Figure 101. A common fossil, yet one of the more exotic members of the plant community in terms of modern distribution, is *Koelreuteria allenii*, the golden-rain tree (Family Sapindaceae). It is a plant that illustrates the uniqueness of the ancient Florissant community, in showing that many living relatives of the fossils are widely dispersed throughout the world and no longer overlap in distribution. Today, *Koelreuteria* is native only to China, Taiwan, and Fiji. The fruit was a three-sided bladderlike capsule that was easily wind-borne for seed dispersal. This specimen represents one side of a fruit capsule with an attached seed on the inner surface. The original fruit was three-dimensional, but it broke into three pieces before it was fossilized. Specimen UCM-34187, courtesy of the University of Colorado Museum.

Scale: × 2.6

Figure 102. This specimen shows the basal portion of a large compound leaf of *Koelreuteria allenii*, the golden-rain tree (Family Sapindaceae). The entire specimen represents a single leaf that is doubly compound (i.e., there are two orders of division within the compound hierarchy), a condition also seen in modern *Koelreuteria* leaves. Specimen USNM-40581, courtesy of the National Museum of Natural History.

Scale: × 0.9

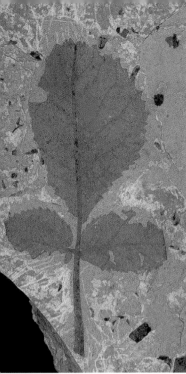

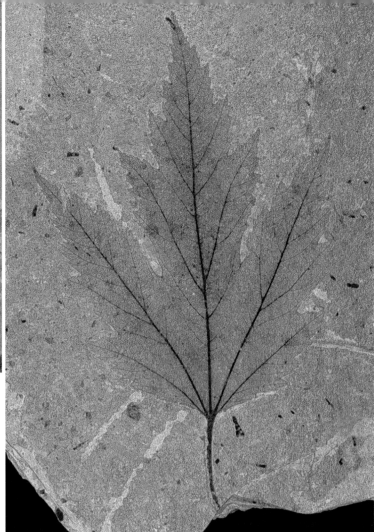

Figure 103. *(above)* This compound leaf with three leaflets is described as *Thouinia striaciata* (Family Sapindaceae). Today, *Thouinia* (guasimilla roja) is restricted to Mexico and the West Indies, where it grows along stream drainages in the upper elevations of the tropical deciduous forest near the boundary with pine-oak woodland. Specimen UCM-38014 (holotype), courtesy of the University of Colorado Museum.

Scale: × 1.2

Figure 104. *(above right)* This perfect leaf of *Acer florissantii* is one of two types of maple at Florissant. Specimen UCMP-3827, courtesy of the University of California Museum of Paleontology.

Scale: natural size

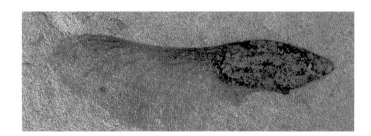

Figure 105. *Acer macginitiei* is a type of maple fruit related to the box elder. The leaves of modern members of this group are typically compound, but they have not been found in the fossil record at Florissant. Winged fruits such as this are excellent candidates for fossilization because they are easily carried by the wind into basins of deposition. Specimen UCMP-3828 (holotype), courtesy of the University of California Museum of Paleontology.

Scale: × 3.3

live in the warmer regions of the southern United States today. *Thouinia* occurs as compound leaves at Florissant (Figure 103), and its modern distribution is limited to Mexico and the West Indies.

Some botanists have suggested that the maple family, Aceraceae, should be included in the family Sapindaceae. *Acer* (maple) is represented by two species at Florissant. *Acer florissantii* is the more common of the two and includes both

Figure 106. This attractive winged fruit is similar to *Ptelea* (Family Rutaceae) the hoptree that is widespread in North America today. The venation of these fruits differs slightly from modern *Ptelea*, however, and the fossils may instead represent another, possibly extinct genus. The darkened area in the center is the seed within the fruit. Specimen NHM-V.18539, courtesy of The Natural History Museum, London.

Scale: × 3.4

Figure 107. This fossil fruit of *Ailanthus americana* (Family Simarouba-ceae) is related to the tree of heaven that today grows in eastern and southern Asia and northern Australia. Many of the plants from Florissant now live only in distant regions of the world, demonstrating that the ancient Florissant community was composed of organisms that no longer live together. Specimen UCM-18714, courtesy of the University of Colorado Museum.

Scale: × 3.5

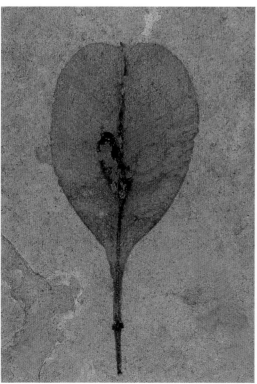
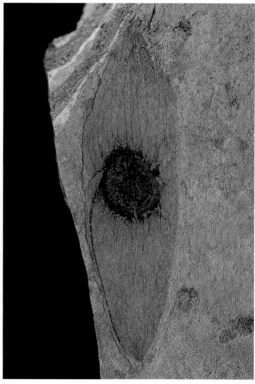

Figure 108. *Cedrela lancifolia* (Family Meliaceae) was probably a tall tree similar to those that grow today in Mexico and Central America, although these leaves are also similar to the closely related genus *Toona*, which now grows in Asia. The leaves were compound as they are in the living species, but as fossils, they are most often found as isolated leaflets such as this. The identification is also confirmed by fossils of the winged seeds. Specimen USNM-39657, courtesy of the National Museum of Natural History.

Scale: × 0.9

leaves (Figure 104) and fruits. Even though they are not found in attachment, both organs are placed into the same species based on comparative inferences from similar modern maples. *Acer florissantii* is classified in an extinct section of the genus. The other Florissant fossil maple is *Acer macginitiei*, known only from winged fruits (Figure 105). These fruits are similar to the modern box elders, which represent a group of maples that some taxonomists suggest should belong in a separate genus.

The presence of the family Rutaceae is indicated by fruits that appear quite similar to those of *Ptelea* (hoptree). MacGinitie described fruits of this genus, but they were later shown to belong to the extinct genus *Diplodipelta* in the family Caprifoliaceae. Other fruits recognized later, however, actually do appear to be *Ptelea* (Figure 106). Its leaves also occur as fossils and are compound with three leaflets. Today, this genus is a shrub or small tree that is native to North America from Canada to Mexico.

The family Simaroubaceae includes *Ailanthus* (tree of heaven), which is confirmed at Florissant from its distinctive winged fruits (Figure 107). The fruits hung in dense clusters at the ends of the branches. *Ailanthus* is one of the plants that today is restricted to areas of the world far from Florissant, occurring only in eastern and southern Asia and northern Australia. It is a tall deciduous tree that easily sprouts from sucker shoots.

The family Meliaceace is represented by *Cedrela* (West Indies cedar) and *Trichilia* (piocha). *Cedrela* occurs both as fossil leaves (Figure 108) and winged seeds. It is a tall deciduous or evergreen tree that grows today in Mexico and Central America and is primarily tropical in distribution, although it is found in other fossil floras that clearly represent temperate deciduous forest. The wood, sometimes referred to as a "cedar," is insect resistant. *Trichilia* also grows in Mexico and Central America and is a shrub or small tree found along watercourses in the tropical deciduous forests and oak woodlands. Its modern distribution also ranges into Africa.

The family Burseraceae is questionably represented by fossil leaves of *Bursera* (elephant tree or torote), but the presence of this tree is also suggested by the fossil pollen record. Modern *Bursera* includes many species of trees that inhabit tropical and subtropical environments in Mexico and Central America. The presence of *Bursera* fossils has been cited as evidence for a warm late Eocene climate at Florissant, based on the nearest-living-relative method of estimating ancient temperatures (see "Reconstructing the Ancient Ecosystem"), because the genus grows only in warm climates today.

Figure 109. The compound leaf of *Rhus stellariaefolia* (Family Anacardiaceae) is one of four Florissant species in this genus, which includes the sumacs and poison ivy. This specimen was once illustrated on a postcard in a series on "Tertiary Fossil Plants" issued by the British Museum during the early 1920s. It is one of many fine specimens collected by Wilmatte Cockerell, who often accompanied her husband, Professor T. D. A. Cockerell, during his collecting expeditions to Florissant in 1906 through 1908. Cockerell was from the University of Colorado, but he exchanged numerous specimens with other museums. As with many of Cockerell's fossils, the two halves of this one ended up in different museums. This specimen is at The Natural History Museum, London, but the corresponding counterpart is still at the University of Colorado. Specimen NHM-V.11399, courtesy of The Natural History Museum, London.

Scale: × 0.75

Figure 110. The leaflets of *Rhus lesquereuxi* (Family Anacardiaceae) have large teeth, which help to distinguish this species from *Rhus stellariaefolia* (Figure 109). Some of the modern species of *Rhus* have leaves with toxins that can be irritating to skin, but there is no way to tell for certain from the fossil record when such characteristics first evolved. Specimen USNM-40787, courtesy of the National Museum of Natural History.

Scale: × 0.8

Figure 111. *Cotinus fraterna* (*below*) (smoke tree) shows secondary veins that branch, sometimes more than once, and extend all the way to the margin, which are characteristics of the family Anacardiaceae. Specimen UCM-18667, courtesy of the University of Colorado Musuem.

Scale: × 1.1

Figure 112. The leaves of *Ziziphus florissantii* (Family Rhamnaceae) have distinct lateral veins and small teeth along the margin. The plant was a probably a thorny shrub or small tree, and its modern relatives are widespread around the world in warm, often dry habitats. The fossil leaves are most similar to living species in Texas and Mexico. Specimen UCM-18634, courtesy of the University of Colorado Museum.

Scale: × 3.5

The family Anacardiaceae (cashew family) is known at Florissant from three genera and six species, consisting of four species of *Rhus* (sumac) and one each of *Cotinus* (smoke tree or chittamwood) and, questionably, *Astronium* (kingwood, locustwood, or urunday). Both fruits and leaves were identified by MacGinitie as *Astronium,* but the fruits were later placed into the extinct genus *Chaneya* (possibly belonging in the family Simaroubaceae), leaving some doubt about the identification of the *Astronium* leaves. *Rhus* is more confidently identified from its compound leaves (Figures 109 and 110), and these plants were small trees, shrubs, or vines. Modern species of this genus turn colorful shades of red during the autumn, and it is likely that the Florissant species did the same. *Rhus* is wide-spread today, and variations in leaf form and venation indicate as many as four species at Florissant. *Cotinus* is known from leaves (Figure 111), and there are only two or three living species of this genus, in eastern North America, southern Europe, and China.

The family Rhamnaceae is represented by *Colubrina* (snakewood), *Rhamnites,* and *Ziziphus* (jujube, ber, or cogwood). *Colubrina* is known from leaves, and the genus is distributed today in warm climates of the world including southern Florida and Mexico. *Rhamnites* is a form genus denoting a type of fossil leaf that has not been placed into a living genus. *Ziziphus* is known from distinctive leaves that are among the smallest from Florissant (Figure 112). *Ziziphus* is widespread today in warm climates around the world, and its nearest occurrence to Florissant is in northern Mexico. Like the living *Ziziphus,* the Florissant species was probably a thorny shrub or small tree that produced edible fruits for wildlife, and its flowers were an attraction for bees.

Hydrangea (Family Hydrangeaceae), like the modern garden varieties of hydrangea, was a shrub with large, showy flower heads. Rarely, the isolated flowers are found as fossils (Figure 113), but the leaves are more common (Figure 114). Both the leaves and flowers are placed into the same species, *Hydrangea fraxinifolia,* even though they have not been found attached. The plants would have grown in moist sites, most likely near streams or by the lake. Another member of this family is *Philadelphus* (mock orange). It is a shrub with

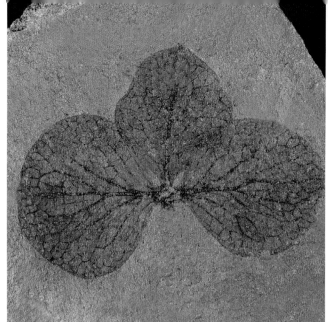

Figure 113. *Hydrangea fraxinifolia* (Family Hydrangeaceae) flowers were probably clustered into large heads, but as fossils, they occur only as rare, isolated calyxes. They are typically four-parted, but one of the sepals is missing in this specimen, probably lost during transport into the lake. Like the modern garden hydrangeas, the flower color could have ranged from white to pink to blue to green, depending in part on the chemistry of the soil. Specimen USNM-33676, courtesy of the National Museum of Natural History.

Scale: × 3.8

Figure 114. This leaf, described as *Hydrangea fraxinifolia*, is thought to correspond to the flowers of *Hydrangea fraxinifolia* (Figure 113), although they have not been found in attachment. Specimen UCMP-3682, courtesy of the University of California Museum of Paleontology.

Scale: × 0.65

Figure 115. This fruiting head, belonging in the family Araliaceae, was borne on a long stalk, or pedicel. The radiating structures in the fruits are the styles of the flowers, which persisted as the flowers developed into these fruits. Specimen WC-FL-23, courtesy of the late Paul R. Stewart, Waynesburg College.

Scale: × 1.5

conspicuous flowers but is known at Florissant only from its small leaves with prominent lateral veins arising from the base.

The family Araliaceae is documented from two specimens of fruiting heads (Figure 115), and from poorly preserved fragmentary leaves thought to be *Oreopanax* (mano de león). These lobed leaves are among the largest at Florissant, and they have large pointed teeth and sinuses (between the teeth) that are broadly curved and almost semicircular in shape. The preservation of these leaves is too poor to warrant illustration here. *Oreopanax* includes many modern species, some of which live in tropical areas of the Americas but others that range into pine-oak woodlands in Mexico.

The family Sambucaceae is known from the leaves of *Sambucus* (elderberry). This shrub or small tree produces large, spreading flower clusters that develop into edible berries. These fruits would have provided an important food source for some of the birds at Florissant. The Florissant species of *Sambucus* had compound leaves with at least five leaflets each (Figure 116). The plant probably lived in the rich soil along the banks of the lake or streams.

Figure 117. Two prominent pairs of arching lateral veins characterize the distinctive leaves of *Smilax labidurommae,* the greenbrier. It was either a vine that climbed the trunks of trees and dangled from the branches, or possibly a shrub. UCMP-3684, courtesy of the University of California Museum of Paleontology.

Scale: × 0.8

Figure 116. The elderberry, *Sambucus newtoni* (Family Sambucaceae), has compound leaves, and this specimen shows that there were at least five leaflets. The teeth are more sharply pointed than most other leaves from Florissant, except for *Paracarpinus.* Specimen USNM-40587A, courtesy of the National Museum of Natural History.

Scale: × 0.85

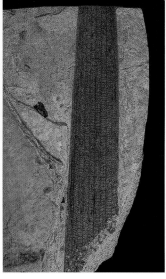

Figure 118. Only these rare fruits of *Stipa florissantii* (Family Poaceae), and rare occurrences of grass pollen, confirm the presence of grasses at Florissant. Widespread grasslands had not yet developed by the late Eocene, and these grasses probably occupied restricted habitats in the drier woodlands away from the lake. USNM-34751 (syntype), courtesy of the National Museum of Natural History.

Scale: × 3.4

Figure 119. *Typha lesquereuxi* was a cattail (Family Typhaceae) that grew in shallow water near the shore of the lake, where these plants formed dense thickets. These straplike leaves also show similarity to *Sparganium* (Family Sparganiaceae). It can be difficult to distinguish these two aquatic genera without flowers or the leaf's cuticle, neither of which is preserved. UCMP-3645, courtesy of the University of California Museum of Paleontology.

Scale: × 1.2

Figure 120. Delicately preserved, this specimen shows the calyx, petals, and stamens of a flower that bears the name *Phenanthera petalifera*. Its taxonomic affinity remains uncertain. Because they are so fragile and do not easily detach from a plant, flowers are rare in the fossil record. Specimen UCM-18594, courtesy of the University of Colorado Museum.

Scale: × 4.7

The families Smilacaceae and Dioscoreaceae are monocotyledonous angiosperms with similar leaves. *Smilax* (greenbrier) was first identified by Cockerell from the leaves at Florissant, and MacGinitie retained the name. Fossil fruits unattached to the leaves, however, show definitive characteristics of *Dioscorea* (yam), casting into doubt whether the leaves actually belong to *Smilax*. Uncertain identifications such as these can sometimes be resolved if the leaves and fruits are found in attachment, or if features other than leaf shape and venation, such as the thin cuticle covering, can be studied, but this has not been possible with these leaves. Nonetheless, the leaves considered to be *Smilax* (Figure 117) are distinctive in both shape and venation, making them easy to distinguish from all other Florissant leaves.

Although widespread grasslands did not develop until long after Florissant's time, fossil fruits recognizable as belonging to the family Poaceae (grasses) do occur here. The presence of grasses is also verified by rare fossil pollen, but stems and leaves have not been found. The fossil fruits (Figure 118) are similar to the modern genus *Stipa*, related to forms that live in oak woodlands today. The Florissant grasses were probably restricted in distribution, growing primarily in the drier, open habitats within the woodlands on the slopes and ridges surrounding the lake basin.

Palms (Family Arecaceae) are rare as leaf fossils at Florissant, and only one or two specimens of a fan palm have been found. Palm also has been confirmed from its fossil pollen. The leaf fossil is too fragmentary and too poorly preserved to be able to identify it to genus, but the presence of palm is important from the standpoint of reconstructing Florissant's ancient climate. Because palms are sensitive to frost, they do not survive if the mean annual temperature is less than 10°C, or if the mean cold month temperature is less than 5°C. Like some of the modern palms, the Florissant species may have been a low shrub or small tree in an open part of the forest or woodland.

Cattails (Family Typhaceae) are aquatic plants that grow in shallow waters, and their abundance in the fossil record at Florissant may indicate nearshore, shallow water conditions at the sites where they are common. Fossils of *Typha lesquereuxii* are fairly common at some sites because the plants grew in dense clusters and needed virtually no transport for their remains to reach the sediments at the lake's bottom. Only the straplike leaves have been found (Figure 119), and the presence of closely spaced cross-veins casts some doubt on the validity of the identification of these as *Typha*. None of the characteristic flower spikes of this genus have been found at Florissant.

The fossil plants shown in Appendix 1 are classified according to their most recent treatment in the paleontological literature. Some of these identifications are more confidently based than others. In the most recent evaluation of the Florissant plants, Manchester noted that the validity of some identifications is doubtful or in need of closer scrutiny. Some of these unconfirmed identifications include plants such as *Astronium*, *Bursera*, *Oreopanax*, and *Thouinia*, all of which have been thought to suggest an affinity of Florissant with the warm modern flora of northeastern Mexico. Some of Florissant's plant fossils still remain to be classified taxonomically, including some of the well-preserved flowers (Figure 120).

BLOWING IN THE WIND: THE MICROSCOPIC WORLD OF POLLEN AND SPORES

Pollen and spores are dispersed by the wind, or carried by insects, from one plant to another during the process of pollination. In the ancient environment at Florissant, some of the pollen was blown into the waters of Lake Florissant, or fell on the surrounding land surface where rainwater later washed it down the slopes and into the lake. Here, the pollen and spores settled to the bottom and were deposited in the fine sediments to become a part of the shale. These microfossils of pollen and spores can be extracted in a laboratory by dissolving the shale in a series of acids. Pollen grains, which have walls composed of a tough organic substance called sporopollenin, survive this treatment in acids and remain as a residue in the test tube after the rest of the shale has been dissolved. The pollen is then stained and mounted on a slide for microscopic examination and is identified by characters such as size, shape, surface sculpturing, and the number and arrangement of furrows and pores. Pollen grains are so small that they are measured in micrometers (indicated as μm, equal to 0.001 millimeters) and range in size from about 20 to 150 μm in diameter.

Figure 121. Palynologists are divided in the way that they name pollen grains. Some use conventional botanical taxonomic names, whereas others use names that describe pollen morphologies. Wingate and Nichols described this pollen as *Tricolpopollenites parmularius*. That name is derived from a type of morphology known as "tricolporate" (tri = three; colpate = grooves; porate = pores), describing the fact that it has three grooves with one pore in the middle of each groove. Two of the grooves are prominent in this photograph, and the other is less distinct. Other palynologists attribute this same type of pollen to the botanical genus *Eucommia* (hard-rubber tree), which is a plant living only in China today. The red coloration is a stain that is added during laboratory preparation of the pollen. Longest dimension 28 μm. Photograph by Hugh Wingate, provided courtesy of Doug Nichols.

Most of the pollen at Florissant is from wind-pollinated plants such as pines and members of the elm, beech, and walnut families. These plants cast their pollen into the atmosphere, releasing many times the amount of pollen as plants that are insect pollinated. The insect-pollinated plants, by contrast, are more conservative in the amount of pollen that they produce, and this pollen is carried only by specific insects. These insects often have coevolved with the plants and are attracted only to certain species, transporting the pollen on their wings or legs to other flowers of the same species. Because of this very specific mechanism of pollen dispersal, and the differences in the amount of pollen produced, insect-dispersed pollen is less likely than wind-carried pollen to reach a lake. And for that reason, wind-pollinated plants are much better represented in the fossil record.

For several reasons, the names that are given to pollen and spores may be different from the names given to leaves and fruits. Pollen and spores usually cannot be distinguished to the level of species and often not to the level of genus. Some palynologists use generic names that are created to describe particular forms of pollen and spores, rather than the conventional generic names that plant taxonomists use. For example, Wingate and Nichols used the fossil pollen name *Tricolpopollenites parmularius* for Florissant pollen (Figure 121) that has characteristics in common with the modern genus *Eucommia*. The name *Tricolpopollenites* denotes the tricolporate nature of this type of pollen; that is, it has three furrows or "grooves" between the poles of the pollen grain, and each furrow has a pore within it. This approach to identification may frustrate a plant taxonomist, but it is a system that developed to satisfy the needs of stratigraphic palynologists, who usually use pollen to determine the age of rocks, especially in oil exploration. Other palynologists, including Leopold and Clay-Poole, prefer to identify the pollen to the same generic categories that are used in plant taxonomy, and these authors identified the same type of pollen from Florissant using the modern generic name *Eucommia*. Because both approaches have been used to name Florissant pollen and spores, two separate lists are given in Appendix 1. Comparisons between the two are not always straightforward.

According to the work of Wingate and Nichols, the Florissant pollen and spore record is quite diverse and contains more than 130 species, including about 100 angiosperms, 18 ferns, 9 gymnosperms, and 6 algae. More than 75 percent of the pollen represents only five families: Pinaceae (pine family), Cupressaceae (cypress family, including the former Taxodiaceae, or redwood family), Ulmaceae (elm family), Juglandaceae (walnut family), and Fagaceae (beech family). The work of Leopold and Clay-Poole recognizes 150 types of pollen and spores, dominated by the same five families.

The record of pollen and spores at Florissant represents a mixture of plants from aquatic, lakeside, slope, and upland habitats. Some of these plants are preserved only as pollen, whereas others are present also as fossil leaves or fruits. Still other plants, known only from leaves or fruits, are absent in the pollen record.

A comparison of Florissant's fossil record for pollen and spores with the record for leaves and fruits clearly reveals that not all plants are preserved both ways. The work of Wingate and Nichols shows that 12 families recognized from leaves and fruits are not represented as pollen and spores, and an additional 5 families are only questionably represented. By contrast, at least 20 families are represented exclusively from pollen and spores. Nineteen families are confirmed from both pollen and leaves/fruits. Similarly, Leopold and Clay-Poole's work shows that 23 of the leaf and fruit genera are corroborated by pollen and spore evidence, and that another 25 genera are known exclusively from pollen and spores. The reason for the differences in the preservation of leaves and fruits versus pollen and spores lies in taphonomy. For example, some plants known only from pollen and spores, such as the family Chenopodiaceae (saltbush or goosefoot family), were small shrubs growing in dry habitats, and their leaves and fruits were not transported to the lake although their wind-blown pollen was. By contrast, plants known exclusively as leaves or fruits, and absent as pollen and spores, include some that were insect pollinated (such as several genera in the rose family), and consequently their pollen did not reach the lake.

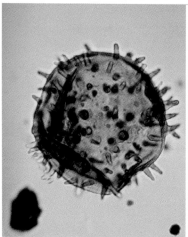 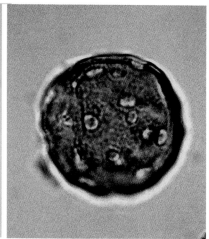

Figure 122. This large, spherical type of pollen with coarse spines is rare at Florissant. Identified as ?*Nupharipollenites*, it belongs to the family Nymphaeaceae (water lily family), a group of aquatic herbs with rounded leaves and large flowers that float on the surface of lakes and ponds. The flower was insect pollinated, and this probably explains why the pollen is so rare in the fossil record, in spite of the fact that the plants lived in the lake. Diameter 53μm (including spines). Photograph by Hugh Wingate, provided courtesy of Doug Nichols.

Figure 123. This type of pollen with many pores is typical of the families Chenopodiaceae and Amaranthaceae, although it is often difficult to make more precise identifications at the generic level. Both of these families usually live in arid or salty habitats today. The record for these families at Florissant comes exclusively from pollen. Diameter 25 μm. Photograph by Hugh Wingate, provided courtesy of Doug Nichols.

Some of the plants known exclusively from pollen and spores provide interesting additions to the fossil flora. We know from the spore evidence, for example, that in addition to the one fern family that is represented by fronds (Aspleniaceae), there were three or four other families of ferns at Florissant (Osmundaceae, Salviniaceae, Schizaceae, and Polypodiaceae [=Aspleniaceae?]). The pollen record also adds three conifers: *Pseudotsuga* or *Larix* (Douglas fir or larch), *Cedrus*-type (perhaps from an extinct relative of the true cedar), and *Tsuga* (hemlock). These were probably conifers from the nearby uplands. The apparent presence of the family Asteraceae (sunflower family) is indicated by its characteristic spiny pollen, and its occurrence at Florissant may represent the earliest known fossil record for this family. It is a family that evolution has since favored, and today Asteraceae is one of the largest plant families in the world.

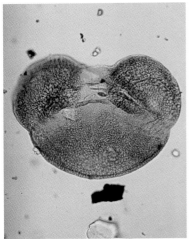 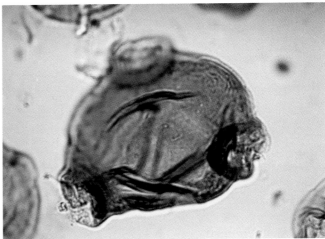

Figure 124. Pollen confirms the presence of *Abies* (fir), a genus that is otherwise known only from one specimen of a seed at Florissant. The pollen is more common than seeds because it was carried longer distances by the wind, and the two large bladders on the pollen grain evolved as an adaptation for wind dispersal. It is likely that fir was confined to cooler habitats far from the lake, such as the volcanic highlands to the west, and the pollen was carried by the wind or water over these long distances. Longest dimension 128 µm. Photograph by Hugh Wingate, provided courtesy of Doug Nichols.

Figure 125. This type of pollen has three large pores and is identified by two names: the morphological name *Corsinipollenites oculus-noctis parvus*, and the botanical name *Xylonagra*. It belongs in the family Onagraceae (evening primrose family) and is similar to the genus *Xylonagra* that grows today only in Baja California, Mexico. Florissant represents the earliest record for the evening primrose family in North America. The pollen is rare because the plant was pollinated by insects. Equatorial diameter 40 µm (longest dimension 53 µm). Photograph by Hugh Wingate, provided courtesy of Doug Nichols.

The pollen of Asteraceae, however, is rare and could represent a modern contaminant that entered the sample in the laboratory during preparation, casting doubt on the validity of its fossil record here. The only representative of the family Solanaceae, *Datura* (jimson weed), was probably a low sprawling vine with large bell-shaped flowers, and its modern relatives are adapted to dry conditions and include hallucinogenic plants used in rituals by the ancient Aztecs and other tribes. The family Rhoipteleaceae is represented by only one modern genus, which lives today in Southeast Asia, and the occurrence of this family at Florissant is apparently a first in the Eocene pollen record. Pollen with distinct spines (Figure 122) indicates the presence of the family Nymphaeaceae

Figure 126. The pollen that corresponds to the fossil foliage and cones of *Sequoia affinis* (redwood) shows some variability at Florissant and is placed in two morphological pollen genera: *Taxodiaceae-pollenites* (left) and *Sequoiapollenites* (right). The two morphological names reflect the two distinct forms of this pollen, even though both types were probably produced by *Sequoia*. *Sequoiapollenites* is most similar to the pollen of living *Sequoia* (coast redwood) and *Sequoia-dendron* (Sierra redwood or giant sequoia) in the presence of a protruding papilla (photograph at right). *Taxodiaceae-pollenites* is the more common type of pollen at Florissant, but it lacks the characteristic papilla. Evidently, the Florissant redwood differed from its modern relatives by producing greater quantities of pollen lacking a papilla. This is yet one more feature to show that the Florissant redwood was distinct from its modern relatives in California. The pollen of *Chamaecyparis*, however, is also similar to that of *Taxodiaceae-pollenites*. Diameters 22 and 24 μm. Photograph by Hugh Wingate, provided courtesy of Doug Nichols.

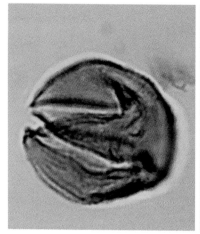
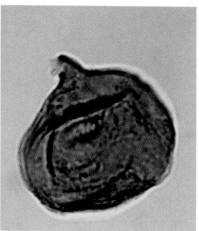

(water lilies), which are aquatic plants. A type of multipored pollen is characteristic of the families Chenopodiaceae and Amaranthaceae (Figure 123), which typically live today in dry or salty habitats, suggesting that similar microhabitats existed in the area around Lake Florissant. Other plant families that are known at Florissant exclusively as pollen (some from Wingate and Nichols's list of pollen morphotype names) include Alangiaceae, Bombacaceae (kapok family; typically tropical to subtropical today), Buxaceae (box family), Cercidiphyllaceae (katsura family; only in eastern Asia today), Elaeagnaceae (oleaster family), Ericaceae (heath family), Malvaceae (mallow family), Polemoniaceae (phlox family), Polygonaceae (knotweed family), and Sapotaceae (chicle family; typically tropical today).

Pollen helps confirm the presence of some plants that are otherwise known only from rare or scanty evidence from leaves or fruits at Florissant. For example, only one seed of *Abies* (fir) has ever been documented, but *Abies* is present in most of the pollen samples (Figure 124). Palm is known only from one or two leaf fragments but is confirmed by the presence of its pollen. Likewise, the grasses, known from a few fruit specimens, are better represented in the pollen record. *Ephedra* (Mormon tea) is known from a single stem, but is verified by the common occurrence of its pollen. *Ephedra* represents a primitive group of plants that today live in dry habitats. *Eucommia* (hard-rubber tree), recently discovered as a fossil fruit, is confirmed by the presence of its pollen (Figure 121). And the rare occurrence of its pollen with large protruding pores (Figure 125) also verifies the identification of the Onagraceae (evening primrose family), known otherwise only from a single flower. All of these examples serve to show how the two different records for fossil plants—pollen and spores, and leaves and fruits—can work together to supplement or support our understanding of what the total ancient flora was really like.

Pollen evidence also can provide insights into evolutionary changes. For example, two different types of pollen (referred by Wingate and Nichols to the pollen genera *Taxodiaceapollenites* and *Sequoiapollenites*) were produced by the Florissant redwood, which is also known from leaves and cones described as *Sequoia affinis* and from *Sequoia*-like stumps. Both types of pollen were found in the fossil male cones of *Sequoia affinis* in a sample prepared by Leopold. These two types of pollen (Figure 126) also occur in modern *Sequoia sempervirens* (coast redwood), but the relative frequencies of the two types are different in the fossil and modern species. In addition to what we have already seen for the wood, cones, and foliage, this is further evidence that the Florissant redwood was in many ways distinctly different from its modern relatives.

SUSPENDED IN THE WATER: THE MICROSCOPIC WORLD OF DIATOMS

Diatoms are microscopic algae that live in water, and they are abundant as microfossils at Florissant (Figures 31 and 127). They have distinctive shapes that range from cigar-shaped to drum-shaped to cylindrical. The cell walls are composed of silica and consist of two overlapping halves that fit together like a hat box or petri dish. These siliceous "shells," or frustules, are porous and ornamented, features important in aiding identification. As photosynthetic algae, diatoms enrich the water with oxygen and are primary producers in the food chain, serving as an important food source for other aquatic organisms. Some diatom species are planktic, living by floating in the water, whereas others are benthic bottom dwellers. Although both planktic and benthic forms are found as fossils at Florissant, benthic forms predominate, and the planktic forms probably were adapted for growth in shallow water. Many represent forms that are probably from sites where the water was less than two meters deep. Diatoms periodically bloom in huge numbers when nutrients are abundant. Assuming that the Eocene forms behaved like modern species, the diatom blooms in Lake Florissant probably covered aquatic plants with a yellow-brown slime and imparted a similar coloration to the lake waters.

As a group, diatoms are common in the fossil record because their siliceous frustules are easily preserved and accumulate to form sediment. Diatoms grow in especially large numbers in areas of recent volcanism, as was the case at Florissant, because silicon-rich volcanic ash provides an essential nutrient for diatom growth. As they grow, many diatoms secrete slimy mucilaginous threads, which cling to those of other diatoms and form large mats, and, as we have already seen, these were important agents in the process of exceptional fossil preservation at Florissant.

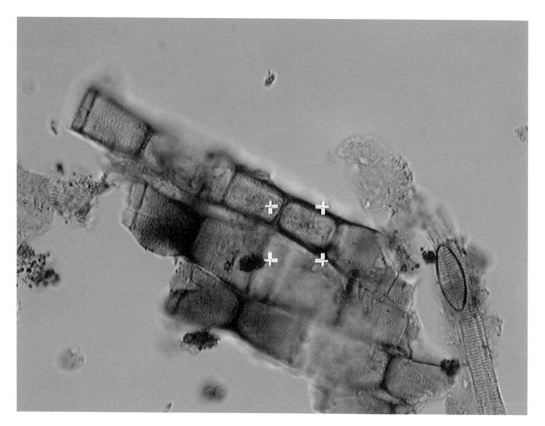

Figure 127. Diatoms are microscopic algae, and many genera are known to have lived in Lake Florissant. Florissant is one of the oldest fossil records for freshwater diatoms. The large cylindrical specimens in this photograph are *Aulacoseira*, which is the most common type of diatom in this sample. The elongate, needle-shaped specimens are *Synedra*, and the smaller elliptical specimen above it is either *Navicula* or *Achnanthidium*. The frame box in the center of the photograph measures 15 µm on a side (inside). Sample from Jess DeBusk; photograph and description courtesy of E. F. Stoermer.

Florissant is significant in representing one of the earliest known occurrences of freshwater diatoms. By contrast, marine forms are known to have existed in the oceans since at least the Jurassic Period. The diatoms known from Florissant show that freshwater forms were already diverse at this time, raising fascinating questions about their evolutionary radiation into freshwater environments. One of the astonishing things about the Florissant diatoms is that they are so diverse and look so similar to modern forms. The high diversity suggests that they may have radiated in freshwater environments well before

Florissant's time. It is thought that modern diatoms disperse from place to place when mists of water are swept up and carried in the wind, and in part, this may be the means by which marine diatoms made the evolutionary transition into freshwater environments.

What we know about Florissant's diatoms is still limited. Little has been published, and much work remains to be done before we have a complete understanding of their taxonomic diversity and adaptation to the freshwater environment. Because of the paucity of information to date, a complete diatom list is not included in Appendix 1. Preliminary examinations by E. Stoermer indicate the presence of at least 19 genera of freshwater diatoms, and it is likely that there are many more. Some of these are probably extinct genera, although many show similarities to living genera that live in shallow, productive bodies of water. Some of these modern forms are planktic, but the majority grow attached to aquatic plants or the substrate. The most abundant types from Florissant are similar to the modern genera *Aulacoseira* and *Synedra,* and less common types are similar to *Achnanthidium, Planothidium, Diatoma, Melosira, Navicula, Eunotia, Gomphonema, Pinnularia, Meridion, Nitzschia, Fragilaria, Staurosirella, Fragilariforma, Ellerbeckia, Epithemia, Rhopalodia,* and *Tetracylus.* Still to be unraveled from these many diatoms are, almost certainly, the important clues about both the early evolution of freshwater diatoms and the aquatic paleoecology of ancient Lake Florissant.

RELATED REFERENCES

Florissant fossil wood: Andrews (1936); Gregory-Wodzicki (2001); Wheeler (2001)

Florissant fossil leaves, fruits, and seeds: Cockerell (1908h); Kirchner (1898); Knowlton (1916); Lesquereux (1883); MacGinitie (1953); Manchester (1989, 1992, 2001); Manchester and Crane (1983, 1987)

Florissant fossil pollen and spores: Leopold and Clay-Poole (2001); Wingate and Nichols (2001)

Taxonomy and ecology of related modern plants: Elias (1980); Felger et al. (2001); Hora (1980); Krussman (1985); Mabberley (1997); Takhtajan (1997)

FOSSIL SPIDERS, INSECTS, AND OTHER INVERTEBRATES

The invertebrate fossils at Florissant consist of arthropods including spiders, millipedes, insects, and ostracods; and mollusks including clams and snails. Of these, the spiders and insects stand out as the most significant. Just as in modern environments, the spiders and insects are a hugely diverse group, with more than 1,500 species described from Florissant. Because of the unusual taphonomic conditions required for their preservation, spiders and insects are much less common than plants in the world's fossil record, making their occurrence at Florissant even more significant. Fossil plant sites abound, but insect sites are rare. The abundance of insects at Florissant results partly from the unusual conditions of taphonomy, facilitated largely by the presence of diatom mats, which provided a rare mechanism of entrapment favoring insect preservation.

Over the past century, the study of fossil spiders and insects has progressed much more slowly than for fossil plants, especially in North America. This is partly because fossil insect sites are so uncommon, but also because there have been far fewer paleoentomologists than paleobotanists. Many of the pitfalls that paleobotanists encountered decades ago are now just beginning to confront paleoentomologists. For example, the validity of most generic identifications of Florissant spiders and insects has not been reevaluated in a century or more. Early workers such as Scudder were "splitters," dividing the different forms within genera into many different species based on minor and in some cases unimportant characteristics. Some of these species need to be combined by synonymy (see "History of Research and Conservation," p. 12). When such studies are undertaken, the more than 1,500 described species of Florissant spiders and

Left: Detail of a superbly preserved female wolf spider. Courtesy of the American Museum of Natural History.

insects are likely to become compressed into a smaller number of valid species, but at the same time, the description of new species may add to the list. The taxonomic list given in Appendix 1 updates higher taxonomic categories, such as families and orders, into contemporary concepts of taxonomy. Approximately 280 genera of the Florissant fossil insects and spiders have been described as extinct, although many of these need to be closely reexamined. Sorting out which genera are correctly identified, however, and how many valid species each genus contains, will be a tremendous task for years into the future. Many genera of fossil insects from Florissant and elsewhere were compiled by Carpenter in his 1992 two volume contribution on fossil insects for the *Treatise on Invertebrate Paleontology*.

Many of the original identifications of fossils were biased by the collections of modern insects that were available for study at the time. Most of the Florissant fossil insects were identified as genera still alive in North America and Europe, but this may simply reflect the fact that during the late 1800s, when the fossils were described, it was those continents that had produced the largest collections of modern insects for comparison. Lesquereux had a similar bias in his identifications of the fossil plants, but those were corrected decades later by MacGinitie's revisions in 1953. A thorough revision, similar to MacGinitie's treatment of the fossil plants, is desperately needed for the insects. In the meantime, we are left to rely on a huge body of scattered taxonomic work more than a century old.

Unlike in fossil plants, the different parts of a spider or insect are usually preserved in attachment, making it easier to reconstruct the entire organism. Occasionally a detached wing is found, but fossil insects typically consist of entire bodies. Paleoentomologists face another problem, however, in attempting to associate the various life cycle stages (i.e., nymphs, larvae, pupae, and adults) of an insect species. The isolated fossils representing different developmental stages present problems similar to those that paleobotanists face when trying to reconstruct entire plants from unattached organs. As fossils, some insects are more abundantly represented by adults than larvae, or vice versa, and it can be nearly impossible to know which fossil larva corresponds to which adult. Furthermore, trace fossils, such as larval cases or leaf-feeding traces, are not easily associated with the insects that made them, and therefore they are given separate names.

Just as modern botanists rely on reproductive features such as flowers as the basis for classification, so, too, do modern entomologists base much of the classification of insects on characters of the genitalia. But genitalia are not preserved in the fossil record at Florissant. Entomologists and botanists who work only with living organisms may be frustrated by the lack of diagnostic reproductive features in the fossils. Paleontologists must work with what they have, however, and this has provided the incentive to look for other characteristics, such as leaf and insect-wing venation, that are also diagnostic in classification.

Aspects of the behavior and habitats of the Florissant insects and spiders must be inferred from their modern relatives, although it is not always possible to know for sure whether the late Eocene insects were necessarily the same, or whether the modern groups may have evolved different lifestyles or adaptations to tolerate different conditions. Nevertheless, the fossil spiders and insects provide many examples that illustrate the complexity of the ancient Florissant ecosystem. They lived in a wide variety of habitats, ranging from strictly aquatic to semiaquatic to fully terrestrial. They include various forms that lived on the ground, in the air, in leaf litter, in trees or herbs, under dead logs, in mammal dung, on the surface of the lake, and in the water. Many of the insects were herbivorous, eating the nutrients produced by plants. Some of these herbivores were detrimental to their plant hosts. Many spiders and insects were predators, feeding on other insects or on the aquatic life in the lake; often they preyed on the herbivorous insects, thus forming a natural food chain. Some of these predators also benefited the plants by controlling the pests that plagued the vegetation. Each of Florissant's plants, spiders, and insects reveals a link in an ancient food chain in what was once a thriving, interactive ecosystem.

We will look at the spiders and insects primarily at the level of their major orders, a higher ranking in taxonomic hierarchy than the level of family that was emphasized for the plants. In a sense, taxonomy shifts by one level in the classification of spiders and insects compared with plants, at least in terms of what we easily recognize. Although plants have features that make it easy to distinguish them to the level of genus, the features of insects are often more difficult to differentiate below the level of family, and therefore the degree of confidence for fossil insect identifications is sometimes more questionable at the generic level. This also is reflected by common names of familiar groups, which typically apply to *genera* of plants, and to *families* of insects. As with the plants, we will look only at selected examples of the major groups of spiders and insects at Florissant. The complete list of Florissant spiders and insects is given in Appendix 1.

Finally, this chapter will examine the other invertebrates at Florissant, specifically the ostracods, clams, and snails. Although certainly not diverse at Florissant, these groups were important members of the aquatic community, and they hold some of the important clues about water conditions in the ancient lake.

ARACHNIDS: THE SPIDERS AND THEIR RELATIVES

The class Arachnida (spiders, harvestmen, mites, ticks, and scorpions) is a group of arthropods that is well represented at Florissant, mostly by the fossils of a large variety of spiders, but also by three species of harvestmen and a questionable gall mite. Arachnids differ from the insects in having eight instead of six legs, two instead of three body regions, and in the lack of antennae and wings, among other features. Their first pair of appendages, the chelicerae, function

Figure 128. Spiders are a large and diverse group at Florissant, although sometimes it is difficult to classify them confidently into higher taxonomic categories. This female of *Eodiplurina cockerelli* is classified as a small tarantula (Order Araneae, Family Theraphosidae). Modern tarantulas are relatively long lived, nocturnal spiders that inhabit burrows by day and hunt by night. This view is the ventral (bottom) side of the spider. Note that the abdomen appears indistinctly between the two hind legs. Specimen UCM-17703 (holotype), courtesy of the University of Colorado Museum.

Scale: × 2.2

Figure 129. This female of *Segestria secessa* is a segestriid spider (Order Araneae, Family Segestriidae). Like modern members of the family, it probably lived in a retreat under rocks or beneath bark. Specimen MCZ-71 (syntype), courtesy of the Museum of Comparative Zoology, Harvard University.

Scale: × 3.9

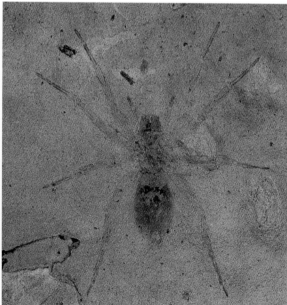

FOSSIL SPIDERS, INSECTS, AND OTHER INVERTEBRATES 131

Figure 130. *Palaeodrassus ingenuus* represents an extinct genus of hunting spider (Order Araneae, Family Gnaphosidae). Like living members of this group, this female may have constructed a tunnel-like retreat under the forest litter or beneath rocks and hunted her prey by night. Specimen MCZ-83 (lectotype), courtesy of the Museum of Comparative Zoology, Harvard University.

Scale: × 3.4

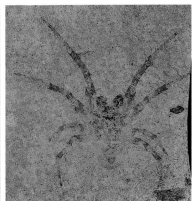

Figure 131. *Tethneus hentzii* was an orb web spider (Order Araneae, Family Araneidae) Orb web spiders capture prey by constructing webs with strands that radiate from the center and are reinforced with circular strands. The enlarged pedipalps in this specimen distinguish it as a male. Specimen MCZ-4145 (syntype), courtesy of the Museum of Comparative Zoology, Harvard University.

Scale: × 2.5

as feeding organs used for piercing and sometimes in attack or defense. The second pair, the pedipalps, are adapted for various functions, such as grasping prey, courtship, and transfer of sperm.

The order Araneae (spiders) is by far the most diverse group of arachnids at Florissant, and several families are represented (Appendix 1; Figures 128–133). Spiders do not undergo metamorphosis, and most hatch from their eggs to begin life looking much like small adults, molting several times as they mature. They have a thin waist between the two body regions, and specialized chelicerae that are adapted to handle prey and inject venom. The spiders are silk producers, and many spiders use this silk to construct webs that aid in the capture of prey. Webs are woven in several basic styles, and each family or genus builds a distinct type of web. The most elaborate web is that of the orb weavers, which build webs with strands that radiate from the center and are reinforced with strands encircling the center. The actual webs are not known as fossils, although orb-weaving spiders have been identi-

Figure 132. With long extended legs, this large female silk spider (Order Araneae, Family Araneidae) probably once walked in an enormous, entangling web that she created in the ancient Florissant forest. Although webs are not known as fossils at Florissant, we do know that modern silk spiders spin strong threads of silk as they build webs a meter or more in diameter. This specimen, identified as *Nephila pennatipes,* measures about 5 cm across. Specimen MCZ-61 (holotype), courtesy of the Museum of Comparative Zoology, Harvard University.

Scale: × 1.2

Figure 133. This superbly preserved specimen is *Lycosa florissanti,* a female wolf spider (Order Araneae, Family Lycosidae). It probably did not spin a web, but lived instead in the ground, under rocks, among the forest litter, or in low herbaceous plants. Like its modern relatives, it may have had patchy, camouflaging colors to remain inconspicuous as it hunted for insect prey, which it seized by pouncing. Specimen AMNH-FI-19032A (holotype), courtesy of the Department of Invertebrate Paleontology, American Museum of Natural History.

Scale: × 2.8

fied from Florissant (Figure 131). These spiders may have spent most of their lives in their webs or in retreats adjacent to the webs, whereas other Florissant spiders were free roaming and hunted along the ground or in shrubs and trees, surprising and capturing their prey. They captured living insects or other spiders, and in turn, many of them were the prey of some of Florissant's larger insects, such as the spider-hunting wasp (Figure 204). Like their modern relatives, Florissant's spider populations were probably very dense, and the various species were partitioned into layered habitats, with some preferring the ground, others inhabiting the layers in herbs and shrubs above the ground, and still others confined to the trees. Some of the spiders occupied specific niches within a habitat, whereas others were more wide ranging.

The most important work on the Florissant spiders was done in 1922 by Petrunkevitch, who reinvestigated many of the fossils originally described by Scudder. As Petrunkevitch noted, it is often easier to name a new species of fossil spider than it is to classify it taxonomically to family. Indeed, there is limited consensus even on the higher-level classification of modern spiders. The difficulties are even greater in the fossils, which must be classified only according to gross external features, rather than the more important microscopic characteristics. Moreover, many of the fossil spiders appear only as indistinct impressions barely discernable from the surrounding rock. This poor preservation is probably due to the soft bodies of spiders relative to the better-preserved insects. For these reasons, the taxonomic assignments of Florissant's fossil spiders must be considered tenuous at best.

One interesting aspect about Florissant's spiders is the way in which the legs are preserved. Spider legs typically curl under the body when a spider dies, unless the water is warm or acidic. The Florissant spiders have their legs extended, and this characteristic has led some to suggest that the waters of Lake Florissant were warm or acidic, perhaps because of volcanic ashfalls or thermal vents related to the volcanism. This interpretation remains speculative, however, and other taphonomic factors may be at play.

Appearing similar to spiders, the order Opiliones (harvestmen or daddy longlegs) can be distinguished by the lack of a thin waist between the body regions. The body is typically round or oval. Harvestmen feed on dead animals or the fluids of plants. Three species of this group are present at Florissant (Figure 134).

The presence of the order Acari (mites and ticks) at Florissant is conjectural. No body fossils have been identified, and only one trace fossil of a gall mite, consisting of several galls on a leaf (Figure 43), has been attributed to this order. Gall mites are plant sucking parasites, and although it is possible that they

Figure 134. Related to the spiders, harvestmen such as *Amauropilio atavus* (Order Opiliones, Family Leiobunidae) have a generally rounded or oval body and long legs. They are rare at Florissant. NHM-I.8427 (holotype), courtesy of The Natural History Museum, London.

Scale: × 2.1

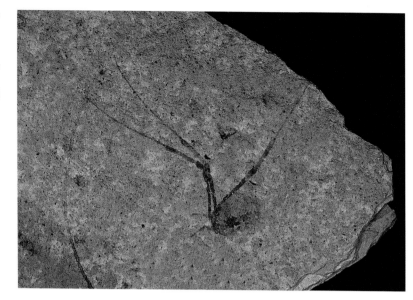

were responsible for the formation of these galls, various types of insects could have formed similar galls.

MYRIAPODS: THE ARTHROPODS WITH MANY LEGS

The myriapods are a group of multi-legged arthropods including the centipedes and millipedes. At Florissant, the group is represented by the class Diplopoda (millipedes) and includes two families, each with a single species (Figure 135). Millipedes look like worms but actually consist of numerous segments, each bearing two pairs of legs. They live in moist habitats in the ground or under leaves, and their unclosable breathing organs and lack of a cuticle covering make them much more susceptible than insects to desiccation. Their diet consists mostly of dead plant material such as decaying leaves, and they were important in recycling nutrients in the ancient ecosystem. They lay their eggs in the summer and spend the winter as adults. The reclusive lifestyle of millipedes explains their rarity in the fossil record.

INSECTS

The insects belong to the class Hexapoda, literally meaning six legs. Insects have antennae as sensory organs, and many have wings during at least a portion of their life cycle. The body is differentiated into three regions: the head, the thorax, and the abdomen. The mouthparts are variously adapted for biting, chewing, piercing, or sucking. Insects lay eggs and go through a complex life cycle,

FOSSIL SPIDERS, INSECTS, AND OTHER INVERTEBRATES   135

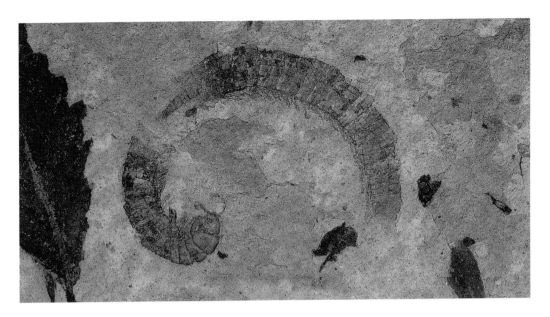

Figure 135. Millipedes (Class Diplopoda) are rare as fossils, and only two specimens have been recovered from Florissant. In both of these fossils, the body is preserved in a somewhat coiled position. The body of a millipede consists of a long series of segments, and each segment bears two pairs of legs. These numerous legs are clearly evident along the inside of this specimen of *Parajulus cockerelli*. Specimen AMNH-FI-22564 (holotype), courtesy of the Department of Invertebrate Paleontology, American Museum of Natural History.

Scale: × 4.3

with the more primitive groups undergoing incomplete metamorphosis and the more advanced groups undergoing complete metamorphosis. Incomplete metamorphosis involves an immature stage of nymphs, which look like small adults without wings or reproductive organs. They molt several times to become adults, finally emerging with wings. Complete metamorphosis involves an immature stage of larvae, such as maggots and caterpillars, and they are heavy feeders as they go through several molts before reaching the last larval stage. They then go into a pupal stage as they transform into adults. Although the adult stage is the most commonly represented in the fossil record at Florissant, in some groups, the adults are actually short-lived. Some groups of insects are highly socialized and colonial and have different castes, with particular body forms, that perform specialized functions within the colony.

The Ephemeroptera (mayflies) is an order of medium-sized insects with long, thin tails and wings that are always held in a vertical position above the body. This inability of mayflies to fold their wings is one of the characteristics that places them among the most primitive groups of insects. Fossils of mayflies are not common at Florissant, although both adults and nymphs are known (Figures 136 and 137), and several species have been described. They lived near the lake and streams, and the nymphs were aquatic. When the nymphs transformed into winged adults, they would rise to the surface of the water and molt (usually twice within about a day), emerging in huge numbers. Although the nymphs lived in the water for a year or more, the adult mayflies lived only for a couple of days, so short a time that feeding was unnecessary. During this short-

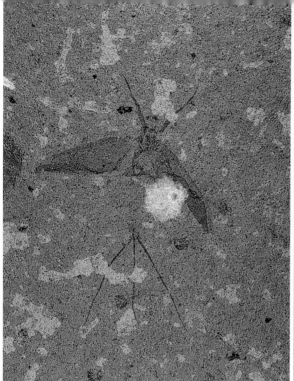 

Figure 136. Mayflies (Order Ephemeroptera) emerge from the water to become winged adults, but this phase of their life cycle only lasts for a couple of days. This short existence of the adult stage gives the order its name, "Ephemeroptera," literally meaning short-lived wing. They returned to the water's surface to lay their eggs, which may be when this adult died and the process of fossilization began. This specimen of *Ephemera howarthi* preserves the three long caudal filaments (tails) and shows the triangular shape of the forewings typical of mayflies. Specimen AMNH-FI-18829A (syntype), courtesy of the Department of Invertebrate Paleontology, American Museum of Natural History.

Scale: × 2.5

Figure 137. This specimen is a mayfly nymph (Order Ephemeroptera, *Ephemera pumicosa*). Most of a mayfly's life is spent in this aquatic nymph stage, which lasts one to two years. This nymph probably lived on the bottom of the lake, perhaps in a shallow burrow. The tail helped it to move about as it fed on algae and other detritus. This fossil is indistinctly preserved because the nymph had a soft body. Specimen MCZ-254 (syntype), courtesy of the Museum of Comparative Zoology, Harvard University.

Scale: × 3.2

lived adult phase, the insects would join in swarming flights around the lake as they mated. The eggs were laid in the water, and the life cycle repeated itself.

The order Odonata consists of two major groups, the dragonflies and the damselflies. They are one of the most primitive groups of flying insects, and very large forms are known from the late Paleozoic Era. Dragonflies and damselflies appear as both nymphs and adults in the fossil record at Florissant (Figures 138–141). The nymphs were entirely aquatic and breathed through gills. They

 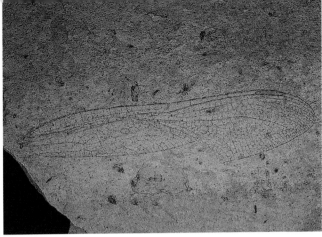

Figure 138. *(above)* Although not well preserved, this specimen of *Aeschna larvata* is the only dragonfly (Order Odonata) nymph described from Florissant. Dragonfly nymphs are aquatic and breathe through gills located within the rectum, drawing water in and out of the anus. This individual was probably capable of rapid movement by shooting water out of its anus. This undoubtedly facilitated its attack on prey, which consisted of various small aquatic organisms. Specimen MCZ-399 (holotype), courtesy of the Museum of Comparative Zoology, Harvard University.

Scale: × 1.8

Figure 139. The intricate patterns of wing venation are the most important characteristics for classifying the various families in the order Odonata, even among living members of the group. Such features make it easy to identify well-preserved fossils, such as this front wing of the adult dragonfly *Hoplonoaeschna separata*. The irregularity along the margin is due to a break in the wing before fossilization. No complete bodies of adult dragonflies have been recorded at Florissant. Specimen MCZ-398 (holotype), courtesy of the Museum of Comparative Zoology, Harvard University.

Scale: × 1.9

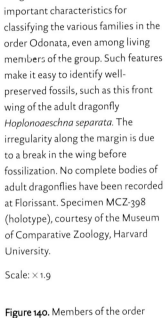

Figure 140. Members of the order Odonata are aquatic during the nymph stage of their development. This damselfly nymph, *Calopteryx telluris*, breathed through the gills, which appear as tail-like appendages. The nymph was able to swim by undulating its body, and the gills functioned similar to the tail of a fish. Specimen MCZ-396 (holotype), courtesy of the Museum of Comparative Zoology, Harvard University.

Scale: × 2.4

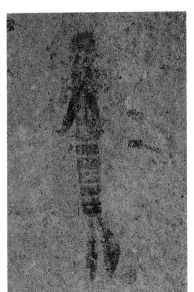

Figure 141. The two major groups within the order Odonata can be distinguished, in part, by the manner in which the wings are held while the insect is at rest. The dragonflies always hold their wings in a horizontal, outstretched postion, whereas the damselflies, such as this specimen of *Miopodagrion optimum*, hold the wings folded in a vertical position above the body. Specimen UCM-8609 (holotype), courtesy of the University of Colorado Museum.

Scale: × 1.7

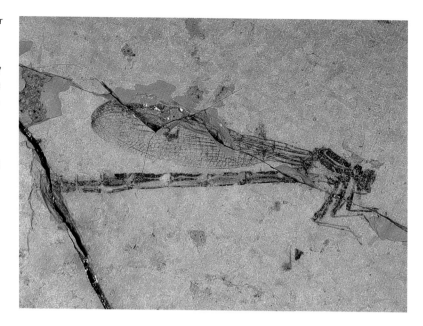

fed on aquatic organisms as large as small fish and were themselves the prey of other aquatic insects and fish. Once fully grown, the nymphs emerged from the water to molt and became adults, which lived for two to eight weeks. These adults were poor fliers at first, but they quickly developed into large, strong-flying insects. They lived near the lake and streams, although some may have ranged into the nearby meadows and woodlands, and they returned to the water to mate and lay their eggs. Like their modern relatives, the adults were probably quite colorful, but any such coloration is not preserved in the fossils. The adults preyed on a variety of insects which they caught while in flight.

The large wings in the Odonata occur in pairs with a hind wing and a forewing. It is the only insect order in which the two sets of wings move independently of one another, with the front pair of wings down while the rear pair is up. The venation in the wings is netlike with many branches, which is a primitive characteristic among insects. It is primarily these patterns of wing venation that are used to distinguish the major families of Odonata, and these characteristics can be well preserved in the fossils. The wings of one individual, however, often are superimposed over one another during fossilization. The manner in which the wings are held while the insect is at rest is one feature that distinguishes the dragonflies from the damselflies. Dragonflies hold the wings in an outstretched and fixed horizontal position, another primitive character among insects. By contrast, damselflies hold their wings folded into a vertical position above the body.

The order Orthoptera includes the grasshoppers, crickets, and katydids,

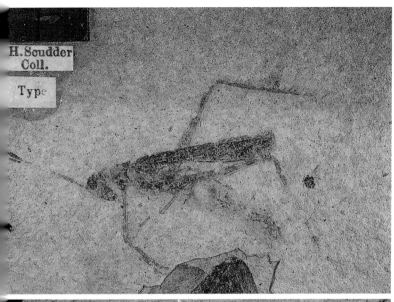

Figure 142. *Tyrbula russelli* is a short-horned grasshopper (Order Orthoptera, Family Acrididae) that probably lived in open meadows during the summer and fall. The fossil plant record confirms the presence of grasses at Florissant, suggesting that the coevolution of grasshoppers and grasses was already well underway by the late Eocene. Modern members of this group make a buzzing sound by rubbing the legs and wings together. Specimen MCZ-7 (syntype), courtesy of the Museum of Comparative Zoology, Harvard University.

Scale: × 2.3

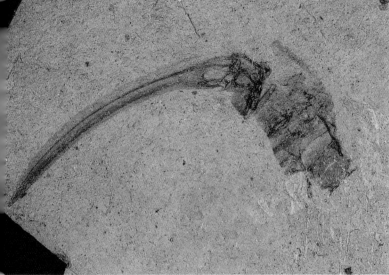

Figure 143. This specimen preserves the egg-laying female ovipositor of *Anabrus caudelli*, a long-horned grasshopper (Order Orthoptera, Family Tettigoniidae). The eggs were most likely laid into plant tissues, where the grasshoppers overwintered in dormancy during this egg stage. Specimen UCM-4536 (holotype), courtesy of the University of Colorado Museum.

Scale: × 4.2

all of which are represented at Florissant (Figures 142–145). They probably filled the air with loud sounds much as their relatives do today, singing by rubbing their legs and wings together. Several different types of grasshoppers have been described from Florissant, although crickets and katydids are rare. Like their modern relatives, they probably had a wide range of different lifestyles, some living in open habitats and forming large swarms, and others living solitary and elusive lives underground or in trees. The adults lived during the warm season, and late in the season, the females used their ovipositors (Figure 143) to deposit eggs either in the soil or in plant tissues. The eggs over-wintered and later emerged as nymphs

Figure 144. This katydid (Order Orthoptera) has not been described yet and may be the only specimen of its kind from Florissant. It is quite large, measuring 6 cm in length. Many members of the Orthoptera typically sing, and katydids do so by rubbing the veins of their wings together. Ones such as this probably filled the air with sound during the late Eocene evenings at Florissant. Specimen FLFO-112, Florissant Fossil Beds National Monument.

Scale: × 1.7

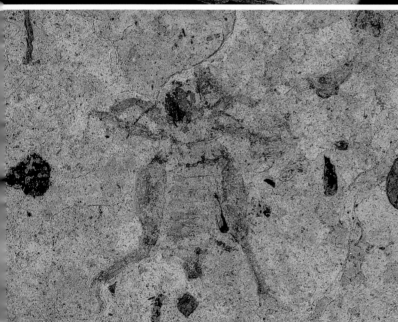

Figure 145. Crickets (Order Orthoptera, Family Gryllidae) are rare at Florissant, and this specimen of *Lithogryllites lutzii* is the only specimen of the only described species. Like its modern relatives, this cricket probably spent its days concealed under rocks or in the ground but emerged during the night. Lifestyles are just one factor that influences taphonomy, and the reclusive existence of crickets may explain in part why they are so rare in the fossil record at Florissant. Specimen UCM-4532 (holotype), courtesy of the University of Colorado Museum.

MScale: × 4.2

that molted several times to become adults during the following warm season. Most of the orthopterans were herbivores, feeding on various types of plants including the grasses that were present in localized habitats. It is unlikely, however, that there were true, widespread grasslands at this time, and many questions about the coevolution of grasses and grasshoppers remain unanswered. When such studies are undertaken, however, Florissant will surely provide an important link.

Figure 146. The mantids (Order Mantodea) are large insects with long necks and large front legs, but at Florissant, the group is represented only by wings, such as this specimen of *Lithophotina floccosa*. Until a complete body fossil is discovered, the presence of mantids remains questionable. UCM-4527 (holotype), courtesy of the University of Colorado Museum..

Scale: × 3.4

The order Mantodea (mantids) is represented only by isolated wings (Figure 146), and the lack of entire bodies makes their presence at Florissant more questionable than some of the other insect groups. Nevertheless, wing fossils, just like isolated leaf fossils, can often be identified by characteristics of venation. Modern mantids are hunters that live in plants and prey on other insects or even small vertebrates, and their movable heads and well-developed front legs are useful for capturing prey. Many of them live in tropical climates, but some range as far north as Canada.

The order Blattaria (cockroaches) is another primitive insect group with a long fossil record extending back to the Pennsylvanian Period. As almost everyone knows, cockroaches are the "rats" of the modern insect world. Cockroaches are rare at Florissant, and only three species have been described (Figure 147). The modern cockroaches vary in their reproductive styles and egg-laying habits, and their diet consists of dead and decaying organic material and feces. They occupy a wide range of habitats but favor warmer climates.

The order Isoptera (termites) is represented at Florissant by five species in three families. Termites evolved from a cockroachlike ancestor, and they are medium-sized, social insects that eat wood. Termites are one of the most primitive insects to form colonies with highly organized castes—including reproductives, workers, and soldiers—each of which is morphologically distinct and performs specific functions in the colony. Large numbers of the winged reproductives develop each year and leave the colony, flying to new locations where they shed the wings, mate, and become the kings and queens of new colonies. Once having established a new colony, these reproductives nourish their young nymphs, many of which in turn develop into wingless workers and soldiers, as

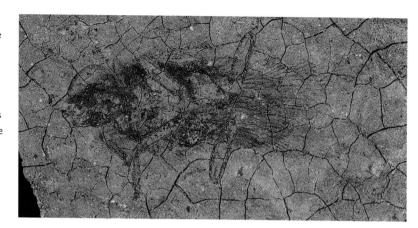

Figure 147. Cockroaches (Order Blattaria) are rare at Florissant, and this specimen of *Zetobora brunneri* is one of only two species. The numerous surface cracks show what happens if the wet, clay-rich Florissant shale dries too rapidly in the sun immediately after being split. Specimen MCZ-470 (holotype), courtesy of the Museum of Comparative Zoology, Harvard University.

Scale: × 3.6

well as a new generation of winged reproductives that repeat the cycle. The workers perform functions such as nest building and feeding, whereas the soldiers defend the nest. Most termites build their nests in close association with dead wood, and some build giant, spirelike mounds. Although examples of these huge mounds have been found in the fossil record as trace fossils, none are known from Florissant. Most of the termite fossils at Florissant consist of reproductives with four distinctive, long wings (Figure 148). The only caste to develop wings, reproductives had a higher preservation potential because they were capable of weak flight and may have been carried by the wind over the lake waters. Termites have softer bodies than similar insects such as ants, and therefore they do not preserve as prominently or abundantly in the fossil record.

Figure 148. Termites (Order Isoptera) at Florissant consist of reproductives, the only one of three castes that produces wings. They are four-winged, but much of the wing membrane is usually indistinct in the fossils. The wings are soon lost after the reproductives fly from their original nest to establish a new colony some distance away. They are not strong fliers, and this may be the taphonomic bias that favored the fossilization of winged reproductives as they flew over the lake. This specimen of *Zootermopsis*(?) *coloradensis* is a dampwood termite (Family Hodotermitidae) that most likely preferred damp, rotting logs. Specimen UCM-4427, courtesy of the University of Colorado Museum.

Scale: × 1.8

FOSSIL SPIDERS, INSECTS, AND OTHER INVERTEBRATES 143

Figure 149. The slender body with long pincers projecting from the tail of the abdomen in this specimen of *Labiduromma gurneyi* is typical of earwigs (Order Dermaptera, Family Labiduridae). The delicate hind wings in this specimen are not visible because they are folded beneath the hardened forewings, which show as the darkened, carbonized area in the thorax of the upper middle part of the body. Specimen UCM-29902 (holotype), courtesy of the University of Colorado Museum.

Scale: × 2.9

The order Dermaptera (earwigs) consists of 12 species at Florissant, all of them placed in the genus *Labiduromma*. They have long, flattened bodies with abdominal pincers. Some of the fossils display the delicate hind wings in an outstretched position, although they are more often folded against the back under short, hardened forewings (Figure 149). The earwigs are mostly nocturnal and spend their days in tight cracks or other confined spaces. Their diet consists of dead and living plants and occasionally other insects. They spend the winter as adults, and the eggs are laid in shallow burrows in the ground where they are tended by the adult females.

The order Embiidina (web-spinners) is not a large group and is represented by only one species at Florissant (Figure 150). Web-spinners live in colonies and construct intricate silk tunnels underground or in the leaf litter. All of the insects in the colony, including the young nymphs, spin silk from glands located

Figure 150. Web-spinners (Order Embiidina) make tunnels of silk underground or in leaf litter. Only the males have wings, such as this specimen of *Lithembia florissantensis*. The dark, carbonized area at the posterior of the abdomen is probably the cercis, which functioned as a sensory organ in the rear. This organ, along with flexible wings that fold forward, allow these insects to move backwards in their tunnels. UCM-4421 (syntype), courtesy of the University of Colorado Museum.

Scale: × 5.5

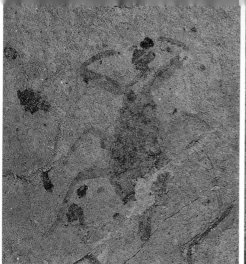 

Figure 151. Several members of the order Hemiptera are aquatic, including the waterscorpions (Family Nepidae). Like the modern waterscorpions, this specimen of *Nepa vulcanica* was predaceous and used the specialized front legs to capture small aquatic animals. A breathing siphon at the rear of the abdomen emerged from the water while the body rested on aquatic plants. Specimen YPM-26166 (holotype), courtesy of Peabody Museum of Natural History, Yale University.

Scale: × 3.5

Figure 152. The backswimmer (Family Notonectidae) is another aquatic member of the order Hemiptera. It swam upside down, on its back, and had specialized rear legs that functioned like powerful oars. This specimen of *Notonecta emersoni* preserves these legs in the forward position, as if the backswimmer were about to take a full paddling stroke. Backswimmers probably lived near the edge of the lake, resting on the water's surface and holding their bodies at an angle with the head down, underwater. They preyed on other aquatic insects or even small fish by using their shorter front legs. Specimen MCZ-3523 (syntype), courtesy of the Museum of Comparative Zoology, Harvard University.

Scale: × 6.8

within the front legs. They undergo incomplete metamorphosis and lay their eggs in the silk tunnels. Their diet consists mostly of plant material. Modern web-spinners live primarily in warm climates.

The order Hemiptera (bugs) is a large group both in the modern world and as fossils at Florissant. Although many insects are commonly called bugs, these are the "true" bugs. They have two sets of wings, and the group is characterized by the structure of the forewings, which are thickened at the base and membranous at the tip. The tips of these forewings overlap when the insects are at rest, sometimes forming an X-shaped pattern on the back. The Hemiptera have piercing and sucking mouthparts within their beaklike rostrum, which is capable of swinging forward from the front of the head. They were important players in the game of plant-

Figure 153. The water striders (Order Hemiptera, Family Gerridae) are semiaquatic, making their living while walking on water. The long legs in this specimen of *Gerris protobates* aided its locomotion as it skated across the quiet surface waters of Lake Florissant. The two shorter front legs (the left one of which is bent back alongside the body) were used to capture prey. Water striders are scarce as fossils at Florissant, which may seem unusual for an insect that actually lived on the lake. They had fine hairs on their legs that helped to keep them buoyant, however, and it is possible that this buoyancy contributed to their scarcity in the fossil record. Specimen NHM-IN.26925 (holotype), courtesy of The Natural History Museum, London.

Scale: × 3.0

Figure 154. Lace bugs (Order Hemiptera, Family Tingidae) are small, and this specimen of *Dictyla veterna* measures little more than 3 mm in length. The ornate patterning on the upper surfaces of the wings is the feature that gives the group its common name. Like its modern relatives, this species was probably a plant feeder that lived on herbaceous plants and trees. Specimen MCZ-662 (syntype), courtesy of the Museum of Comparative Zoology, Harvard University.

Scale: × 9.9

insect interactions. Most of the Florissant bugs were terrestrial plant feeders and used the sucking mouthparts to extract nutrients from the broad-leaved plants in the ancient community. By contrast, bugs that lived underwater or on the water surface were predators, eating aquatic prey or other small insects that fell onto the lake's surface. Some of the terrestrial bugs also were predators. The various families of Hemiptera occupied a wide range of habitats, and examples from Florissant include aquatic bugs such as waterscorpions (Nepidae; Figure 151), water boatmen (Corixidae), and backswimmers (Notonectidae; Figure 152); semiaquatic bugs such as riffle bugs (Veliidae) and water striders (Gerridae; Figure 153); and terrestrial bugs such as lace bugs (Tingidae; Figure 154), leaf bugs (Miridae; Figure 155), assassin bugs (Reduviidae; Figure 156), seed bugs (Lygaeidae; Figure 157), squash bugs (Coreidae), burrower bugs (Cydnidae), and stink bugs (Pentatomidae).

Figure 155. Leaf bugs (Order Hemiptera, Family Miridae) are represented by about a dozen species at Florissant, including this one of *Poecilocapsus fremontii*. Leaf bugs lived on vegetation and probably fed on a variety of different plants. The membranaceous tips of the front wings overlapped when the bug was at rest, forming the X-shaped pattern on the back that is typical of many families in the Hemiptera. Specimen MCZ-650 (syntype), courtesy of the Museum of Comparative Zoology, Harvard University. Photograph by John Fraser.

Scale: × 7.0

Figure 156. The assassin bug (Order Hemiptera, Family Reduviidae) was a terrestrial predator that most likely fed on other insects, although a few of the modern members of this group will suck the blood of mammals and are carriers of disease. This specimen of *Poliosphageus psychrus* shows the development of a "neck," a feature that is often present in assassin bugs. Specimen UCM-18621 (holotype), courtesy of the University of Colorado Museum.

Scale: × 2.8

Figure 157. Seed bugs (Order Hemiptera, Family Lygaeidae) represent the most diverse family in the order Hemiptera at Florissant. Like its modern relatives, this specimen of *Lygaeus stabilitus* probably ate seeds. The antenna to the left shows its typical segmentation into four parts. Specimen PU-6811 (syntype), from the former Princeton University collection, now at the National Museum of Natural History. Photograph by Doris Kneuer.

Scale: × 4.2

The order Homoptera is closely related to the Hemiptera, and is divided into two major groups, one including the cicadas and hoppers, and the other the psyllids, aphids, and scale insects. All of these are present in the fossil record at Florissant (Figures 158–161). They exhibit a variety of body forms, and some go through a rather complex life cycle. Like the Hemiptera, the Homoptera have sucking and piercing mouthparts, but the beak is directed downward instead of forward. Winged members of the group have two pairs of wings, with the forewings being uniform in texture, unlike the Hemiptera. Members of the Homoptera, like those today, probably were exclusively plant feeders in the ancient community at Florissant, and some may have had a significant detrimental impact on certain plants by excessive consumption or by the transmission of disease. Some forms such as aphids may have been responsible for causing blight or extensive deformation of

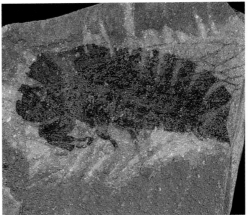

Figure 158. Froghoppers (Order Homoptera, Family Cercopidae) are small hopping insects that feed on plants. The various modern representatives of the group often feed on specific host plants, but it is unknown whether their feeding habits were similarly specialized by the late Eocene, or if so, which plant species they favored. The patterns on the wing suggest that this species of *Palecphora maculata* was two-colored, as are many of the modern froghoppers. Like the modern froghoppers, the nymphs of this species probably secreted and lived in frothy, spitlike masses on leaves and fed from plant saps. Specimen MCZ-5013 (syntype), courtesy of the Museum of Comparative Zoology, Harvard University.

Scale: × 4.4

Figure 159. *Platypedia primigenia* is one of three species of cicada (Order Homoptera, Family Cicadidae) known from Florissant. It was moderately large as an adult, and like the living cicadas, it probably had a life cycle that took as long as several years to complete. The males may have made loud sounds with their abdominal organs during courtship or in response to disturbance. Specimen YPM-26165 (holotype), courtesy of Peabody Museum of Natural History, Yale University.

Scale: × 2.8

Figure 160. *(left)* This specimen of *Florissantia elegans* shows the ventral side of a planthopper (Order Homoptera, Family Dictyopharidae). Modern members of this family feed on grasses, which also were present at Florissant during the late Eocene. *Florissantia* is one of many extinct insect genera at Florissant. Oddly, this is also the generic name for an extinct fossil plant (see Figure 83). Separate rules apply for naming plants and animals, and although rare, it *is* possible that a plant and an animal can have the same Latin name. Specimen PU-6783 (syntype), from the former Princeton University collection, now at the National Museum of Natural History. Photograph by Amanda Cook.

Scale: × 4.3

Figure 161. The aphids are relatively inactive insects that feed extensively on plants. This specimen is placed in the extinct genus and species *Siphonophoroides pennata* (Order Homoptera, Family Drepanosiphidae). Aphids may have been responsible for making some of the leaf galls at Florissant. Specimen MCZ-1031, courtesy of the Museum of Comparative Zoology, Harvard University.

Scale: × 5.8

plant organs in their hosts. Some of the Florissant Homoptera probably fed on a wide variety of plants, whereas others were more specialized. Clearly, the Homoptera represent another important group in the realm of plant-insect interactions and coevolution, but much more work remains to be done before their effects on the ancient Florissant vegetation can be well understood.

The order Neuroptera is represented at Florissant by two of its three suborders, Raphidiodea (snakeflies) and Planipennia (lacewings). Some authorities consider these groups to represent separate orders. Members of the Neuroptera are terrestrial, soft-bodied insects with four membranous wings that are held like a roof above the body when the insects are at rest. They undergo complete metamorphosis and lay their eggs on foliage or bark. The larvae are predaceous, feeding on other small insects. The snakeflies (Figure 162) are distinguished by their long "necks," a feature that evolved by an extension of the front thoracic segment. This long-necked adaptation raises the head above the rest the body, facilitating the capture of prey. Snakeflies are nourished by sucking the fluids from their victims. The lacewings, too, are predatory and may have been beneficial to some of Florissant's plants by feeding on aphids or other herbivorous insects. Some modern lacewings also feed on pollen or nectar. The lacewings had relatively large wings with intricately branched venation, and the forewings and hind wings were nearly equal in size and shape (Figure 163). An exceptional fossil of a thread-winged lacewing has been found (Figure 164). The larvae of lacewings probably pupated within silk cocoons as they do today.

The order Coleoptera (beetles) was by far the most diverse and dominant of any group of organisms in the ancient Florissant community. Beetles account for 40 percent of modern insect species and 38 percent of the insect species known as fossils at Florissant. The beetles occupy a wide range of ecological niches, with habitats ranging from terrestrial to underground to aquatic, and sometimes including specialized habitats such as the interior of tree trunks. They undergo complete metamorphosis including a resting pupal stage. The length of the life cycle, the life stage in which winter is passed, and the form of the larvae all vary considerably from one family of beetles to another. The beetles have mandibles in their mouthparts that are well adapted for chewing. Their feeding habits range from herbivorous to scavenging to predaceous, and their diets range from leaves and wood to dung to other small insects. Many groups of beetles were important in their interaction with plants. Like a few groups of

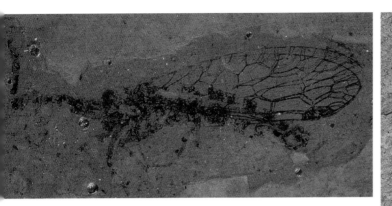

Figure 162. Several species of snakeflies (Order Neuroptera, Family Raphidiidae) have been described from Florissant, including this one of *Raphidia tumulata*. Snakeflies are so named because of the long, necklike prothorax that extends to the front of the body and holds the head. This long-necked adaptation enables the head to be raised above the rest of the body, allowing the snakeflies to strike their prey much like a snake does. Specimen MCZ-4137, courtesy of the Museum of Comparative Zoology, Harvard University.

Scale: × 6.8

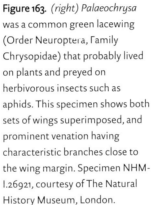

Figure 163. *(right) Palaeochrysa* was a common green lacewing (Order Neuroptera, Family Chrysopidae) that probably lived on plants and preyed on herbivorous insects such as aphids. This specimen shows both sets of wings superimposed, and prominent venation having characteristic branches close to the wing margin. Specimen NHM-I.26921, courtesy of The Natural History Museum, London.

Scale: × 3.4

Figure 164. The thread-winged lacewings (Order Neuroptera, Family Nemopteridae) such as *Marquettia americana* had long, thin hind wings that broadened near the tips. Modern members of the group lay eggs on the ground, where the larvae develop and pupate in a cocoon. This family is widespread in subtropical to tropical areas today but no longer lives in North America. Specimen AMNH-FI-18886 (holotype), courtesy of the Department of Invertebrate Paleontology, American Museum of Natural History.

Scale: natural size

Figure 165. The ground beetles (Order Coleoptera, Family Carabidae) represent a large family both in the modern world and at Florissant. This specimen of *"Plochionus" lesquereuxi* probably lived on the ground, walking through the forest litter by night and hiding under leaves or beneath fallen logs during the day. Judging from modern ground beetles, this one most likely was a hunter that preyed on other insects during the night, and it may have fed on insects that were harmful to plants. The illustration on the left shows the insect's dorsal side, and the one on the right is the ventral side. The lighter areas visible in the elytra on the dorsal side may correspond to what were color variations. Specimens MCZ-1854 and 1855 (holotype), courtesy of the Museum of Comparative Zoology, Harvard University.

Scale: × 6.3

modern beetles, some of the Florissant species probably were capable of producing sound through a variety of bodily motions, as by rubbing together specialized ridges to produce characteristic sounds known as stridulations.

One of the most important diagnostic features of beetles is found in the structure of the wings. Beetles typically have four wings, with a thickened, hardened front pair that covers a membranous hind pair. The front pair are referred to as the elytra, and when folded, they join to form a straight line down the middle of the beetle's back. The elytra form a protective covering over the longer and more delicate hind wings. Because these hind wings normally are only exposed and used during flight, they are seldom evident in the fossils. Hard parts are always favored by taphonomic processes, and the thickened elytra and other hardened body parts of beetles helped to make the Coleoptera better suited for fossilization than many other groups of insects. Even in some beetles of small size, tiny bumps and ridges on the elytra are exceptionally well preserved in the fossils. Because the bodies of beetles are hardened and three-dimensional in character, the differences between dorsal and ventral surfaces is often obvious when the two corresponding halves of a fossil specimen are compared (Figures 35, 165, 166, and 168).

The order Coleoptera is a huge group that is subdivided into a number of suborder, superfamily, and family groups. Although it is possible to illustrate only a few of the important families here (Figures 165–179), the beetles dominate Florissant's taxonomic list (Appendix 1) with about 600 species! In his 1992 compilation of fossil insects for the *Treatise on Invertebrate Paleontology*, Carpenter questioned the validity of many of the generic assignments for the Florissant beetles. Genera of such doubtful assignment are indicated by placing the generic name within quotes, as is done frequently for the beetles in Appendix 1.

 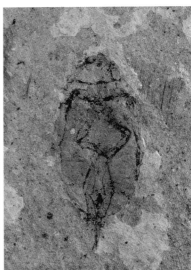

Figure 166. *"Agabus" charon* was a predaceous diving beetle (Order Coleoptera, Family Dytiscidae) that lived in the waters of Lake Florissant. Modern members of this group are aquatic insects, although they are also capable of brief flights onto the land at night. They rise to the surface of the water to breathe, and because they are able to carry air under the elytra, they can remain submerged for a long time. They are capable swimmers, using the powerful hind legs for propulsion. As their name implies, predaceous diving beetles feed on other small aquatic animals, including some as large as small fish. They are similar in form and habitat to the water scavenger beetles (Family Hydrophilidae), and both have been identified from Florissant. The illustration on the left shows the insect's dorsal side, and the one on the right is the ventral side. Specimen UCM-28038 (holotype), courtesy of the University of Colorado Museum.

Scale: × 5.5

Figure 167. The rove beetles (Order Coleoptera, Family Staphylinidae) are characterized by their slender shape and short elytra. The right elytron can be seen just above the legs in this specimen of *"Bledius" osborni*. The abdomen, visible here as the curved, segmented portion of the body, is flexible and mobile and is used to push the hind wings under the elytra. With the abdomen bent upward as in this specimen, rove beetles are capable of running quickly, but they are also fast fliers. Rove beetles occupy a range of habitats including soil, leaf litter, or decaying plants. Specimen PU-6605 (syntype), from the former Princeton University collection, now at the National Museum of Natural History.

Scale: × 16.3

Figure 168. The scarab beetles (Order Coleoptera, Family Scarabaeidae) represent a large group that is divided into many subfamilies. This specimen of *"Macrodactylus" pluto* is a chafer (Subfamily Melolonthinae), a group of plant feeders that eat flowers and foliage. Modern larvae of this group live in the soil, where they eat roots and are harmful to plants. The illustration on the left shows the insect's dorsal side, and the one on the right is the ventral side. Specimen UCM-8255 (holotype), courtesy of the University of Colorado Museum.

Scale: × 4.0

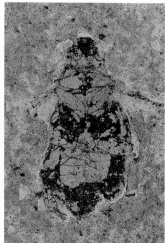

Figure 169. This scarab beetle (Order Coleoptera, Family Scarabaeidae) is an aphodian dung beetle (Subfamily Aphodiinae). Modern members of this group are common in cow dung. This specimen of *"Aphodius" inundatus* most likely inhabited and ate the dung of herbivorous mammals such as the brontotheres, oreodonts, and small horses. Specimen USNM-90441 (holotype), courtesy of the National Museum of Natural History.

Scale: × 6.2

Figure 170. Metallic wood-boring beetles (Order Coleoptera, Family Buprestidae) are brilliantly colored in metallic tones of green, red, and blue. Such coloration is preserved in the fossil record at the Eocene site in Messel, Germany, but not at Florissant. Like its modern relatives, this specimen of *"Chrysobothris" gahani* probably inhabited fallen logs and the foliage of trees. The eggs were probably laid in the bark, and the larvae bored into the wood, leaving a winding maze of tunnels in which the larvae pupated. This specimen shows the two elytra (hardened forewings) spread from the body, exposing one of the membranous hind wings. Specimen UCM-4510 (holotype), courtesy of the University of Colorado Museum.

Scale: × 4.0

Figure 171. The click beetles (Order Coleoptera, Family Elateridae) are named for their ability to click as they jump. They live on plants, leaf litter, or rotting wood. The elytra were slightly separated when this specimen of *"Elater" granulicollis* was preserved. Specimen YPM-1 (holotype), courtesy of Peabody Museum of Natural History, Yale University.

Scale: × 2.0

Figure 172. Soldier beetles (Order Coleoptera, Family Cantharidae) have relatively soft, long bodies. Like its modern relatives, this specimen of *"Podabrus" wheeleri* probably lived on flowers in the meadows or along the edge of the forest, where it preyed on other insects or consumed pollen. Specimen YPM-7 (holotype), courtesy of Peabody Museum of Natural History, Yale University.

Scale: × 3.4

Figure 173. *(left)* Ladybugs or ladybird beetles (Order Coleoptera, Family Coccinellidae) have oval, brightly colored bodies. The light, roughly circular areas that are barely evident on the elytra of this specimen of *"Chilocorus" ulkei* may correspond to what were patterns of different coloration. The two hind wings can be seen protruding at the posterior end of the body from underneath the elytra. These hind wings were used during flight, but normally they were protected beneath the elytra while the insect was at rest. Like the modern ladybirds, this one probably fed on herbivorous arthropods such as aphids and mites. Specimen MCZ-1593 (syntype), courtesy of the Museum of Comparative Zoology, Harvard University.

Scale: × 6.0

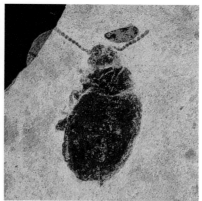

Figure 174. Tumbling flower beetles (Order Coleoptera, Family Mordellidae) typically have an arched body profile and a pointed protrusion at the posterior of the abdomen. This specimen of *"Mordellistena" scudderiana* probably had its niche mostly on flowers, where it ate nectar. Modern relatives in this group prefer the flowers of the sunflower family (Asteraceae), which is one of the largest families of flowering plants. Pollen evidence suggests that the sunflower family makes its first known appearance in the fossil record at Florissant. Could it be that this beetle once sat on one of the world's earliest sunflowers? Specimen MCZ-2693, courtesy of the Museum of Comparative Zoology, Harvard University.

Scale: × 6.5

Figure 175. Darkling beetles (Order Coleoptera, Family Tenebrionidae) feed on dead plant material and live in decaying logs or on the ground. Many of the modern relatives of this group live in desert areas, and it is possible that some of the Florissant members of the group, such as this specimen of *"Platydema" antiquorum,* may have occupied the drier habitats on the hillsides and ridges around Lake Florissant. It is also possible, however, that this group of beetles adapted to dry conditions after Florissant's time. Specimen UCM-8263 (holotype), courtesy of the University of Colorado Museum.

Scale: × 4.8

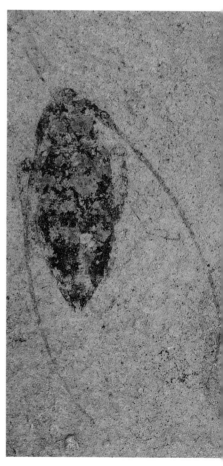

Figure 176. The long-horned beetles (Order Coleoptera, Family Cerambycidae) have long antennae that extend toward the posterior of the body. In many species, the antennae are longer than the insect's body. The larvae of many modern species bore tunnels into trees and are damaging to plants. Florissant species such as this *"Leptostylus" scudderi,* like their living relatives, may have produced larvae that bored into such plants as *Malus* (apple) or members of the elm family (*Cedrelospermum* or *Ulmus*), among others. Specimen MCZ-2598 (holotype), courtesy of the Museum of Comparative Zoology, Harvard University.

Scale: × 5.4

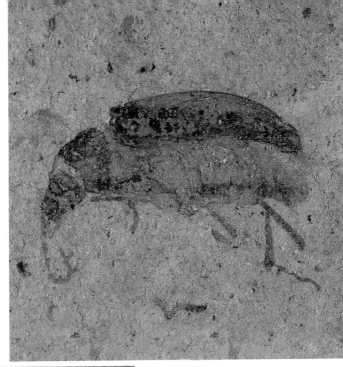

Figure 177. *(right)* The leaf beetles (Order Coleoptera, Family Chrysomelidae) are close relatives of the long-horned beetles but have much shorter antennae. Like its modern relatives, this species of *Colaspis aetatis* would have fed on flowers and foliage. The larvae of leaf beetles may be responsible for some of the leaf mining seen in the fossil leaves at Florissant. Specimen UCM-8219 (holotype), courtesy of the University of Colorado Museum.

Scale: × 9.6

Figure 178. *(left)* Some of the leaf-rolling weevils (Order Coleoptera, Family Attelabidae) are small, such as this specimen of *Trypanorhynchus minutissimus,* which measures only 2.5 mm in length. In spite of its small size, the body and elytra show detailed features. Modern members of this group lay their eggs by cutting slits into a leaf and creating a leaf roll in which a single larva develops. No evidence for this sort of reproductive behavior has yet been discovered in the fossil record at Florissant. The absence of such evidence could mean that this type of behavior had not yet evolved, or that insect fossils such as this one are not validly identified, or that fossil leaves with this type of damage simply have not been found or recognized yet. Specimen USNM-90587E (syntype), courtesy of the National Museum of Natural History.

Scale: × 14.6

Figure 179. The snout beetles (Order Coleoptera, Family Curculionidae), also known as weevils, have long snouts that evolved as an extension of the head. The chewing mouthparts occur at the end of the snout, allowing these weevils to bore holes into the plants on which they feed. As in modern relatives in the Subfamily Curculioninae (nut weevils), the long snout on this specimen of *Curculio restrictus* was probably used to bore holes in which to lay eggs within a nut, where the larvae developed. It may have been a pest on such Florissant plants as the oaks and hickories. The Curculionidae not only contains the largest number of species at Florissant, but it also is one of the largest families among modern members of the animal kingdom. Specimen AMNH-FI-39435A, courtesy of the Department of Invertebrate Paleontology, American Museum of Natural History.

Scale: × 5.2

Members of the order Mecoptera (scorpionflies) have long bodies with four wings of roughly equal size. The males of the family Panorpidae have a long abdomen with an upturned extension bearing the genitalia, forming a structure that superficially resembles the tail of a scorpion (Figure 180). The scorpionflies undergo complete metamorphosis, laying eggs in the soil. The eggs hatch into caterpillar-like larvae, and pupation occurs within a cell underground. Several species are known from Florissant, and they were probably omnivores that fed on either living or dead insects as well as pollen and nectar. They probably lived

Figure 180. Scorpionflies (Order Mecoptera, Family Panorpidae) have two pairs of nearly equal sized wings and an upturned extension of the abdomen that forms a scorpion-like tail. In this specimen of *Holcorpa maculosa*, the tail is folded back over the insect's body. The distinct patterns within the wings were variations in coloration, most likely white areas within a reddish-brown wing. Scorpionflies probably lived in the shady areas within the forest. Specimen AMNH-FI-18887 (allotype), courtesy of the Department of Invertebrate Paleontology, American Museum of Natural History.

Scale: × 2.6

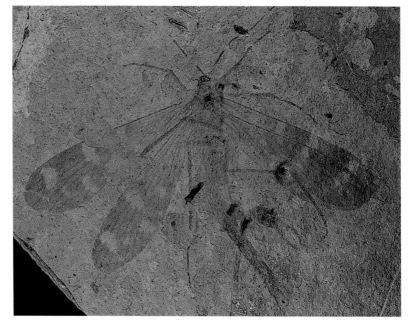

Figure 181. Among the order Diptera, the crane flies (Family Tipulidae) are the most diverse at Florissant. Most of them are large insects, and they have a long slender abdomen, long legs, and like all dipterans, only two wings. The elongated front of the head that protrudes forward of the conspicuous eyes in this specimen of *Tipula heilprini* is typical of the Subfamily Tipulinae. This adult probably only lived for several days and inhabited damp areas in the dense vegetation around the lake. Crane fly larvae also are found as fossils at Florissant. Specimen MCZ-Scudder-11806 (syntype), courtesy of the Museum of Comparative Zoology, Harvard University.

Scale: × 1.9

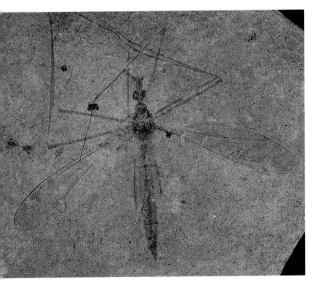

mostly in the moist habitats near the lake. The group has a long history extending back to the late Paleozoic Era, and more species are known from the fossil record than from living members of the order.

Although many unrelated insects bear the common name "fly" as a part of their name, only members of the order Diptera (flies) are the "true" flies. As the name Diptera indicates, flies have only one pair of normal wings, the other pair having become highly reduced into knobbed structures called halteres, which function as sensory organs to stabilize flight. The single set of wings gives flies great maneuverability in flight. Although the two sets of wings in other orders may overlap and become superimposed in their preservation as fossils, you can always be certain that when you see four wings, it is not a true fly. Flies go through complete metamorphosis, and the wormlike larvae are referred to as maggots. The larvae live in diverse habitats, and some are aquatic whereas others live in plant tissues and create leaf mines and galls. The pupae are well protected and adapted to survive extreme conditions. Like flies today, the species at Florissant were important in recycling organic nutrients in the food chain. The maggots fed on decaying plants or animals, whereas adults typically fed on plant nectars, the blood of animals, or other insects. They had mouthparts adapted for piercing, sucking, and lapping. The adults of many species were important pollinators of plants, transporting pollen from flower to flower on their legs or hairy bodies.

The flies at Florissant are a diverse group, just as they are today. Selected examples of some of the larger or more interesting families are illustrated in Figures 181–189. The crane flies (Tipulidae; Figure 181) make up one of the largest families of flies at Florissant, and they lived near the waters of the lake. The larvae of midges (Chironomidae) were aquatic, and their fossils are common in certain layers of the Florissant lake shales (Figure 183). Other important and diverse families at Florissant include robber flies (Asilidae; Figure 184), bee flies (Bombyliidae; Figure 185), dance flies (Empididae; Figure 186), and syrphid flies (Syrphidae; Figure 187). Some of the dipterans were biting flies that possibly transmitted diseases, and the Florissant tsetse flies (Muscidae; Figure 189) represent the only known fossil record for these modern carriers of sleeping sickness. We have no way to know whether such flies were disease carriers in the Eocene, but it is certainly pos-

Figure 182. March flies (Order Diptera, Family Bibionidae) such as *Bibio explanatus* were probably most common in open areas and clearings during the spring and early summer. Modern adult march flies are slow fliers that stay near the ground and frequent flowers. Specimen AMNH-FI-26485 (holotype), courtesy of the Department of Invertebrate Paleontology, American Museum of Natural History.

Scale: × 5.3

Figure 183. Larvae of the nonbiting midges (Order Diptera, Family Chironomidae) are common in some layers of the Florissant shale. These aquatic larvae were important in Lake Florissant's food chain, eating small aquatic plants and animals as well as decaying material, and serving as food for fish and other large aquatic animals. Most of the life cycle of midges was spent in this larval stage. The adults, which lived for only a couple of weeks, also occur as fossils. Specimen FLFO-2532, from the collection of Florissant Fossil Beds National Monument.

Scale: × 2.8

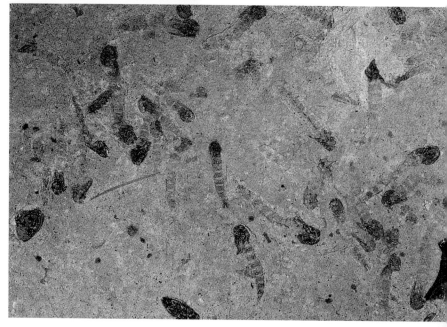

sible. Mosquitoes also belong to the Diptera, and although many novice observers have referred to some of Florissant's fossil insects as "mosquitoes," mosquitoes have not been documented from Florissant.

Adult members of the order Trichoptera (caddisflies) appear similar to moths and lacewings. Fossils of both adults and larvae are present at Florissant (Figures 190–193). They are closely related to the moths, but differ in having hairs instead of scales on the wings. Caddisflies differ from the lacewings by the relative simplicity of the wing venation, which has fewer cross-veins. They have two pairs of wings that fold like a roof over the body. The caddisflies are the only group of insects to have both complete metamorphosis and a larval stage that

Figure 184. Robber flies (Order Diptera, Family Asilidae) can be quite large, and this specimen of *Microstylum wheeleri* measures 4 cm in length. It probably preyed on other insects as large as grasshoppers, dragonflies, and wasps. T. D. A. Cockerell's expeditions to Florissant in 1906–1908 produced one of the largest and best preserved collections ever to come from Florissant, and Cockerell referred to this specimen as "the finest fossil insect found at Florissant by the expedition of 1906." The foliage just above the wing is that of *Sequoia*. Specimen UCM-4522 (holotype), courtesy of the University of Colorado Museum.

Scale: × 1.7

Figure 185. The bee flies (Order Diptera, Family Bombyliidae) are so named because of their superficial resemblance to some of the hymenopterans, but they are easily distinguished because bee flies have only one pair of wings. Like its relatives today, this adult *Alepidophora pealei* probably lived in an open, sunny, dry habitat some distance away from the lake, and it hovered motionless in the air over the flowers from which it fed. Specimen YPM-102 (holotype), courtesy of Peabody Museum of Natural History, Yale University.

Scale: × 4.1

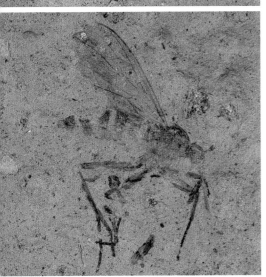

Figure 186. The dance flies (Order Diptera, Family Empididae) most likely lived in the moist, thick vegetation around Lake Florissant. This specimen of *Empis perdita* has a thick proboscis that extends downward from the head and was used as a piercing organ for drinking nectar or the fluids of other small insects. The males of modern species sometimes offer their prey to females as a courtship ritual during mating. Specimen UCM-18637 (holotype), courtesy of the University of Colorado Museum.

Scale: × 5.6

Figure 187. The hover flies (Order Diptera, Family Syrphidae) are moderately diverse in the fossil record at Florissant. They are so-named because of their capability to hover during flight. Judging from its modern relatives, the dark areas in the abdomen of this specimen of *Syrphus aphidopsidis* were black, and the lighter areas probably were yellow, giving this fly a wasplike appearance. This adult probably fed on pollen and nectar from flowers and was an agent in pollination, but the larvae may have fed on aphids just as many of the modern relatives of this genus do. Specimen UCM-8566 (holotype), courtesy of the University of Colorado Museum.

Scale: × 4.0

Figure 188. *Ophyra vetusta* was a small anthomyiid fly (Order Diptera, Family Anthomyiidae) that measured little more than 3 mm in length as an adult. The larvae of this family may have been responsible for forming leaf mines or galls in some of the fossil leaves. Specimen AMNH-FI-26614 (holotype), courtesy of the Department of Invertebrate Paleontology, American Museum of Natural History.

Scale: × 10.1

is aquatic in streams and lakes. One of the interesting aspects of caddisflies is that the aquatic larvae of many species pupate inside cases that they construct in a variety of forms from materials such as sand grains or tiny pebbles, small leaves or twigs, and even the tiny shells of ostracods (Figure 192). The shape and composition of these cases varies between species. Because the larval cases were constructed of durable materials and were used during an insect's aquatic phase, they were good candidates for fossilization in the lake-bottom sediments. Although the larvae can withdraw into their cases for protection, they often move actively with the thorax and legs extended from the opening as they feed on

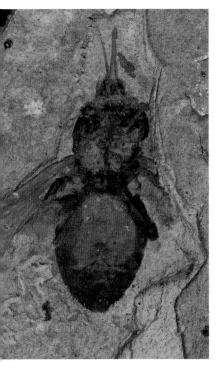

Figure 189. In the modern world, tsetse flies (Order Diptera, Family Muscidae, Subfamily Glossininae) are disease-carrying, blood-sucking insects indigenous to equatorial Africa. Their occurrence at Florissant, including this specimen of *Glossina oligocenus*, is the only known fossil record for this group. The presence of tsetse flies in the late Eocene of North America provides a good example of geographic extirpation and raises intriguing biogeographical questions about the history of their distribution. Modern tsetse flies are carriers of sleeping sickness, and it is possible that these insects plagued the early mammals of Florissant, such as the brontothere and primitive *Mesohippus* horse. Specimen AMNH-FI-45416, courtesy of the Department of Invertebrate Paleontology, American Museum of Natural History.

Scale: × 4.0

Figure 190. Adult caddisflies (Order Trichoptera) hold their wings roof-like above the body. This specimen of *Mesobrochus imbecillus* is thought to be a net-spinning caddisfly (Family Hydropsychidae), and probably lived near streams and laid its eggs in the water, where the larvae developed. Specimen MCZ-319 (syntype), courtesy of the Museum of Comparative Zoology, Harvard University.

Scale: × 6.9

Figure 191. *(left)* This aquatic larva of *Hydropsyche scudderi* is identified as a net-spinning caddisfly (Order Trichoptera, Family Hydropsychidae). The larvae of modern members in this family build a retreat into which they can hide, but they also construct a net in the water, between rocks, for capturing prey from the moving stream currents. The thorax is the curved darkened area at the top, and the long abdomen is the faintly preserved impression extending downward. Specimen NHM-IN.26642 (syntype), courtesy of The Natural History Museum, London.

Scale: × 3.1

Figure 192. This unusual specimen is the name-bearing holotype for two different species, one a caddisfly and the other an ostracod. The overall structure represents the larval case constructed by a caddisfly (Order Trichoptera), described as *Indusia cypridis*. During the aquatic larval stage, many types of caddisflies construct larval cases out of materials such as sand grains, tiny pebbles, pieces of leaves, twigs, or other materials. This species constructed a characteristic type of case using the tiny shells of ostracods along with small, black crystals of biotite mica. Ostracods are microscopic crustaceans, and this specimen contains many individuals, described as *Cypris florissantensis*. Specimen AMNH-FI-35853 (holotype), courtesy of the Department of Invertebrate Paleontology, American Museum of Natural History.

Scale: × 8.5

Figure 193. This specimen preserves the emergence of a winged adult from a larva. It is probably a caddisfly (Order Trichoptera), although it has not been studied in detail and could possibly be a moth (Order Lepidoptera). These two orders are closely related. The identification might be resolved from microscopic examination of the wings to determine whether they have hairs (caddisflies) or scales (moths). Specimen NHM-IN.26250, courtesy of The Natural History Museum, London.

Scale: × 5.5

plants or small animals. The caddisflies overwintered in the water during the larval stage, and the adults emerged during the summer and lived for about a month in terrestrial habitats alongside the lake or by streams. They laid their eggs in or near the water.

The order Lepidoptera includes both the butterflies and moths (Figures 194–198). Florissant is recognized for having more known species of fossil butterflies than any other site in the world. Fossil butterflies attracted special attention from the time they were first discovered at Florissant, perhaps in part because of the early discovery in the 1870s of the perfect specimen of *Prodryas persephone* (Figure 196), which has been renowned ever since. It is probably the finest fossil butterfly in existence. In all, 12 species in 11 genera and three fami-

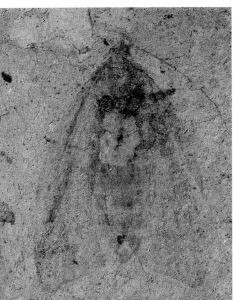

Figure 194. Moths (Order Lepidoptera, Family Tortricidae) are typcially active at night, unlike the butterflies. They are less diverse than butterflies at Florissant, although several species are known, including this *Tortrix florissantana*. Specimen UCM-8579 (holotype), courtesy of the University of Colorado Museum.

Scale: × 3.7

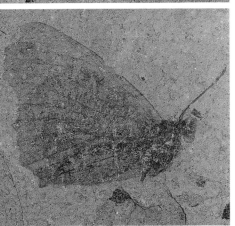

Figure 195. This specimen of *Oligodonta florissantensis* preserves the lateral view of a pierid butterfly (Order Lepidoptera, Family Pieridae). Close examination reveals the coiled proboscis (tongue) beneath the head. The proboscis was an extension of the mouthparts that uncoiled to suck nectar from flowers. The specimen shows little degradation, and some of the characteristic wing scales have been observed microscopically, indicating that the insect probably was preserved soon after its death. The species was described from a specimen that was donated to the National Park Service during the acquisition of the Colorado Petrified Forest, and the counterpart of that specimen, as illustrated here, was later found at Waynesburg College. As with other remarkable fossils, it was not unusual for the two halves of the same specimen to become separated. Specimen WC-FL-1 (counterpart of holotype), courtesy of the late Paul R. Stewart, Waynesburg College.

Scale: × 1.8

Figure 196. Perhaps the world's finest fossil butterfly specimen, *Prodryas persephone* was among the earliest discoveries at Florissant and was described by Scudder in 1878. Scudder referred to it as "in a wonderful state of preservation, the wings expanded as if in readiness for the cabinet and absolutely perfect...." It is a brush-footed butterfly (Order Lepidoptera, Family Nymphalidae) similar to forms that now live in Central and South America. Some of the markings visible in the wing of the fossil represent what were coloration patterns in the living butterfly. The specimen bears catalog number MCZ-1 out of a collection of thousands of fossil insects. Specimen MCZ-1 (holotype), courtesy of the Museum of Comparative Zoology, Harvard University.

Scale: × 2.2

Figure 197. This fossil butterfly was collected in the early 1930s and was once used by tour guides to impress the early tourists who visited Florissant. It was later described as a new species of nymphalid butterfly (Family Lepidoptera, Family Nymphalidae) and became the holotype for *Vanessa amerindica* (Amerindian lady). It is one of only two Florissant fossil butterflies to be placed into a living genus, and *Vanessa* is widely distributed in the modern world. Specimen UF-21999/FLFO-108 (holotype), from the collection of Florissant Fossil Beds National Monument and housed at the Florida Museum of Natural History, University of Florida.

Scale: × 1.8

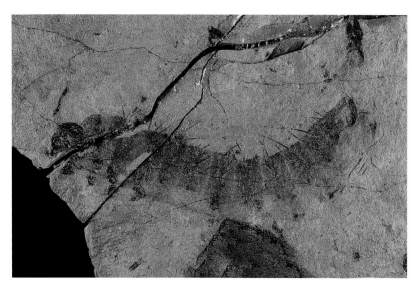

Figure 198. *Phylledestes vorax* was the larva, or caterpillar, of what was probably a noctuid moth (Order Lepidoptera). This specimen clearly preserves the tiny "hairs" around the larva's body. Specimen UCM-4608 (holotype), courtesy of the University of Colorado Museum.

Scale: × 3.4

lies represent Florissant's butterflies. Nine of these species are related to modern groups that occupy the northern New World Tropics. *Prodryas* belongs to the family Nymphalidae (nymphalids or brush-footed butterflies), which is the most diverse family of Lepidoptera at Florissant. Although most of the Florissant butterflies, including *P. persephone,* are placed into extinct genera, at least one of the nymphalids, *Vanessa amerindica* (Figure 197), belongs to a modern genus. Moths also occur at Florissant (Figure 194), but they are less diverse than the butterflies, with about nine species. By contrast, among living Lepidoptera, moth species outnumber butterfly species by at least ten to one. In many respects, the distinction between butterflies and moths is an artificial one, and some of the characters that supposedly distinguish the two actually overlap.

Butterflies and moths are characterized by the presence of a proboscis that extends during feeding and by scales on the wings that give the insects their color. Both of these minute features have been observed in the fossil of *Oligodonta florissantensis,* in the family Pieridae (Figure 195). Lepidopterans undergo complete metamorphosis with a complex life cycle. The larvae of Lepidoptera, commonly referred to as caterpillars, are rare as fossils (Figure 198). The pupa, or chrysalis, may develop within a silk cocoon produced by the larva, or it may be naked. The chrysalis is the stage in which the wingless larva develops into a winged adult. Much of the feeding and growth takes place during the larval stage before this transformation. These larvae undoubtedly fed extensively on the plants at Florissant and were probably responsible for some of the leaf damage seen in the plant fossils, including margin feeding, skeleton feeding, and leaf mining.

The order Hymenoptera (sawflies, wasps, bees, and ants) is another large

Figure 199. Sawflies (Order Hymenoptera, Family Tenthredinidae) may appear superficially similar to the true flies in the Order Diptera, but they are easily distinguished by the presence of four wings instead of only two. Modern sawflies are common in cool temperate climates around the world. They lay their eggs in twigs and leaves on which the larvae later feed, sometimes forming galls or leaf mines as they do. The adults of many species are predators, however. Sawflies spend the winter as pupae, living in cocoons on the ground. Sawflies are well represented by a number of species at Florissant, including this specimen of *Tenthredella fenestralis*. They probably lived in open areas or in woodlands. Specimen NHM-IN.26916 (holotype), courtesy of The Natural History Museum, London.

Scale: × 5.3

group both today and in the fossil record. A selection of fossils representing a few of the many important families of Hymenoptera is illustrated in Figures 199–208. The members of this order have two pairs of wings, the hind wings smaller than the forewings. All, except the group including sawflies (which is the most primitive group of Hymenoptera), have one or two narrow abdominal segments that produce a distinct constriction in the body to form a waist. Metamorphosis is complete in hymenopterans and has many complex variations in the way in which the eggs are laid and the larvae and pupae develop. Comparison with similar modern relatives indicates that many of the Florissant species were parasitic and sought hosts—either plants or the eggs, larvae, or adults of other insects—in which to lay their eggs, providing a place for their own larvae to develop. The ovipositor (egg-laying organ) is often a prominent posterior feature in the Hymenoptera, and it varies from a sawlike structure in the sawflies to a long, thin structure used by the parasitic wasps to penetrate plants and other insects in order to deposit their eggs (Figure 200). In some of the wasps and social ants, the ovipositor of the females has evolved the function of stinging. Many species of Hymenoptera display complex social behavior and are colonial, but others are solitary. The social species such as the ants and bees develop distinct castes, and both reproductives and workers can be distinguished. Among the ants, just like the termites, there was a taphonomic bias that favored the preservation of winged reproductives, because they were capable of weak flight and may have been carried by gusts of wind over the surface of the lake. Many of Florissant's Hymenoptera were attracted to the flowers of particular plants, where they fed on nectar and were important agents in pollination.

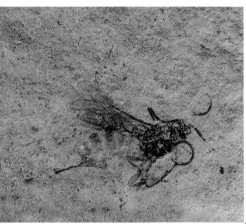

Figure 200. The braconid wasps (Order Hymenoptera, Family Braconidae) are a group of parasitic hymenopterans, laying their eggs either on or inside the eggs, larvae, or adults of other insects. A long ovipositor, such as the one clearly protruding from the posterior of the abdomen in this adult specimen of *Bracon cockerelli*, is used for depositing the eggs. The braconid wasp larvae later develop on or within the host. Several other families of Florissant hymenoptera were similarly parasitic on other insects. Specimen NHM-IN.19224 (paratype), courtesy of The Natural History Museum, London.

Scale: × 8.0

Figure 201. The ichneumon wasps (Order Hymenoptera, Family Ichneumonidae) are a diverse group at Florissant. They were parasitic wasps that laid their eggs on or inside the bodies or eggs of other insects or spiders. The ichneumons had long abdomens, and their long antennae typically became curled before the insect was fossilized, as they are in this specimen of *Parabates memorialis*. Specimen UCM-8575 (holotype), courtesy of the University of Colorado Museum.

Scale: × 2.6

Figure 202. The chalcid wasps (Order Hymenoptera, Family Chalcididae) are a parasitic group that have distinctive, expanded femora on their hind legs. These appendages are clearly evident in this specimen of *Chalcis praevalens*. Specimen AMNH-FI-18898 (holotype), courtesy of the Department of Invertebrate Paleontology, American Museum of Natural History.

Scale: × 5.2

 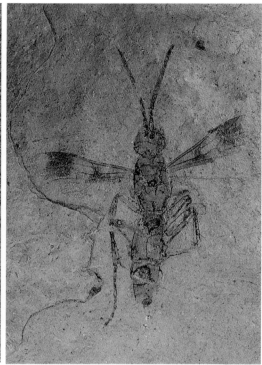

Figure 203. Leaf-cutter bees (Order Hymenoptera, Family Megachilidae), such as this specimen of *Heriades laminarum,* could cut perfectly circular pieces from the edges of leaves (see Figure 45). Judging from their relatives today, these insects were solitary, and they used the small leaf pieces to line the cells of their nests, which were probably made in the cavities of wood. Specimen MCZ-2005 (holotype), courtesy of the Museum of Comparative Zoology, Harvard University. Photograph by April Kinchloe.

Scale: × 6.7

Figure 204. The spider-hunting wasps (Order Hymenoptera, Family Pompilidae) are bold predators that use their long legs to hunt a variety of spiders. Modern members of this group carry a captured, paralyzed spider into a nest cell and lay an egg on the body, and the wasp larva then feeds on the flesh of the spider. The alternating light and dark areas on the wings of this *Salius scudderi* correspond to what were probably color variations. Specimen MCZ-2024 (holotype), courtesy of the Museum of Comparative Zoology, Harvard University.

Scale: × 4.4

Figure 205. The mammoth wasps (Order Hymenoptera, Family Scoliidae), as their name suggests, are large. The modern members of this family hunt and sting scarab beetle larvae, which form the hosts on which the wasps lay their eggs. Once they hatch, the wasp larvae consume their hosts. This specimen represents an extinct genus described as *Floriscolia relicta*. Specimen AMNH-FI-28775 (holotype), courtesy of the Department of Invertebrate Paleontology, American Museum of Natural History.

Scale: × 2.1

Figure 206. This ancient wasp was a yellow jacket or hornet (Order Hymenoptera, Family Vespidae). It was probably a social insect that constructed papery, multichambered nests. This particular specimen of *Palaeovespa* serves as the logo for Florissant Fossil Beds National Monument, and its image has been exhaustively reproduced on clothing, postcards, letterheads, pins, fossil replicas, an entry sign (Figure 16), the outline of a wind vane, and even the cover of this book. Specimen FLFO-50, from the collection of Florissant Fossil Beds National Monument.

Scale: × 3.2

Figure 207. The ants (Order Hymenoptera, Family Formicidae) occur in huge numbers as fossils at Florissant. Ants are highly social insects that are differentiated into castes having different functions and body forms, including both winged reproductives (queens and males) and wingless workers. The winged males and queens emerge in large numbers once a year, and after the mating flight, the males die and the successful queens lose their wings and establish a new colony. The wingless workers make up most of an ant colony, but it is the winged reproductives, such as this specimen of *Protazteca elongata*, that are by far the most common as fossils. Because ants are not strong fliers, it was probably commonplace for them to fall into the water during flight over the lake. This weak ability for flight was the taphonomic bias that favored the fossilization of winged reproductives. Specimen MCZ-2821 (paratype), courtesy of the Museum of Comparative Zoology, Harvard University.

Scale: × 7.1

Figure 208. Worker ants (Order Hymenoptera, Family Formicidae) lack wings, and although they are the most abundant caste in a colony, they are actually uncommon as fossils. Workers such as this specimen of *Archiponera wheeleri* performed the functions of colony construction, food gathering, and protection of the young. Only rarely did their bodies wash into the lake. Ants have one or two segments in their abdomen that are very small, forming a constriction of the body behind the hind legs. Specimen MCZ-2876A (holotype), courtesy of the Museum of Comparative Zoology, Harvard University.

Scale: × 4.2

OSTRACODS: MICROSCOPIC CRUSTACEANS

Members of the class Ostracoda (mussel shrimps) are microscopic freshwater or marine arthropods belonging to the subphylum Crustacea, which is the same group that includes crabs and barnacles. Ostracods have tiny appendages and live within a two-valved, shell-like carapace that encloses them and gives them an appearance similar to minute clams. The carapace opens to enable extension of the appendages during movement or feeding. It is these nearly microscopic carapaces, only about 0.7 millimeters in length, that are found as fossils at Florissant (Figures 192 and 209). The ostracods probably lived on the bottom of the lake and fed on organic matter in the sediment, or on algae. As fossils, they are concentrated in certain layers of the shale (especially in the upper shale unit) but are rare or totally absent in most other layers. Only one species, *Cypris florissantensis,* has been described from Florissant. Ostracods are often good environmental indicators, and future study of the Florissant species may help us to better understand what the water conditions were like in Lake Florissant.

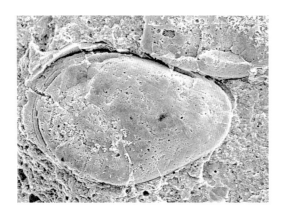

Figure 209. The ostracods (Class Ostracoda) are microscopic crustaceans that are enclosed within two tiny valves and appear almost clamlike. This illustration, taken with scanning electron microscopy, is greatly magnified, and shows one of the two valves (0.7 mm in length). Photograph courtesy of Neal O'Brien.

Figure 210. Freshwater clams and snails are common in some layers of the Florissant shales, but absent in others. The larger shell is the clam *Sphaerium florissantense*, which required well-oxygenated waters to live. Beside it is the coiled shell of the snail *Lymnaea*, which was able to withstand a much wider range of water conditions. These were small organisms, and the clam measures only 9 mm across. Specimen AMNH-FI-18923 (syntype), courtesy of the Department of Invertebrate Paleontology, American Museum of Natural History.

Scale: × 6.6

Figure 211. The shell of this terrestrial snail, *Vitrea fagalis,* coils more than seven times in one plane. Land snails are rare at Florissant, and this specimen may have been carried out into the lake while it was still attached to a leaf. Diameter 7 mm. Specimen UCM-4112, courtesy of the University of Colorado Museum.

Scale: × 9.5

THE MOLLUSKS: CLAMS AND SNAILS

The mollusks include several freshwater and terrestrial types at Florissant. Only one species of freshwater clam, questionably placed in the genus *Sphaerium,* has been described (Figure 210). Like its modern relatives that still inhabit the Rocky Mountains, it was a filter feeder that lived on diatoms and other algae in the lake. *Sphaerium* clams do not tolerate low levels of oxygen, and their presence in Lake Florissant indicates that the waters were well oxygenated. Their fossils, however, occur predominantly in the upper shale unit. This raises questions as to why they are less common in other units of the Florissant Formation, and thus whether there may have been periods during which the lake waters were depleted in oxygen.

The most abundant and diverse group of fossil mollusks at Florissant are the gastropods (snails). Four genera and six species have been described, most of which have relatives that still live in western North America. *Lymnaea,* a genus of freshwater snail with as many as three species at Florissant, is characterized by its high-spired shell (Figure 210). Another freshwater snail, assigned to the genus *Planorbis,* is distinguished from *Lymnaea* by having a shell that is whorled almost in one plane. *Lymnaea* and *Planorbis* were tough organisms that lived by grazing on algae both in turbid streams and in the quiet waters of the lake. They were able to survive a wide range of water conditions and could even survive in wet mud during dry periods. *Lymnaea* and *Planorbis* snails had lungs and rose to the surface of the water to breathe, after which they could remain submerged again for several days. Two land snails, doubtfully assigned to the genera *Omphalina* and *Vitrea* (Figure 211), are also present at Florissant. These also had lungs, and they lived strictly in the terrestrial environment. Like their modern relatives, they probably fed on fungus and mold in the leaf litter.

RELATED REFERENCES
Florissant fossil arachnids: Petrunkevitch (1922); Størmer et al. (1955)
Florissant fossil insects: Brues (1910a); Carpenter (1930, 1992); Cockerell (1907c, 1908g, 1909e, 1911a); Emmel et al. (1992); Melander (1949); Scudder (1890b, 1893, 1900); Wickham (1912b, 1913b, 1914b); Wilson 1978
Florissant fossil mollusks: Cockerell (1906c)
Taxonomy and ecology of related modern insects: Borror et al. (1989); McGavin (2000); O'Toole (1986)

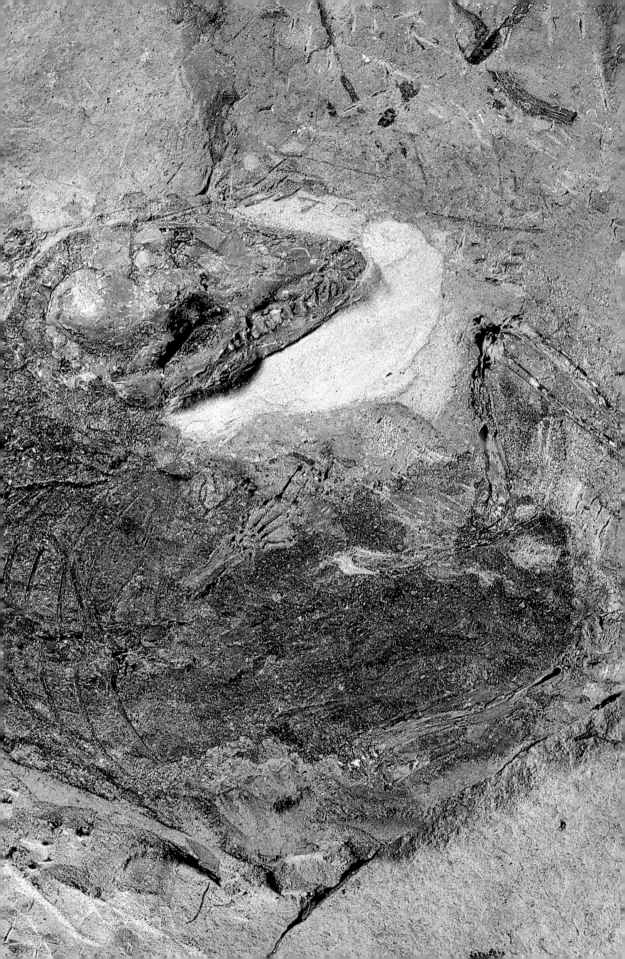

FOSSIL VERTEBRATES

Although large vertebrate fossils may be the attractions that catch the public's eye at many of the world's most famous paleontological sites, Florissant has gained its reputation instead through the huge diversity of plant and insect fossils. Not infrequently, visitors to the site inquire about dinosaurs but are disappointed to learn that these popular extinct animals can have absolutely nothing to do with the label "fossil beds." Dinosaurs, of course, had long expired by the time Florissant was accumulating its fossils.

Vertebrates, except for the fish, are rare and usually only fragmentary in the fossil record at Florissant, and for the most part, we are left to wonder what kinds of vertebrate animals once lived here. The rarity of vertebrate fossils at Florissant is the result of taphonomic biases and the nature of the sedimentary environments in which the fossils were preserved. In part, terrestrial vertebrates are rare in the lake deposits simply because it was uncommon for their dead carcasses to be transported into the lake. In addition, acidity in the lake, caused by decaying plant material or volcanic ashfalls, may have inhibited the fossilization of vertebrates by dissolving bone material. Almost all of the mammal fossils known from Florissant have come from the deposits of the lower mudstone unit (see Figure 36) that formed along the floodplain of the ancient Florissant valley at a time when the lake did not exist, although fossils are uncommon in these rocks. The lake shales have yielded only one small mammal, in addition to several birds and a variety of fish. In spite of the rarity of vertebrate fossils, Florissant nevertheless holds a few important clues for interpreting late Eocene vertebrate life in the area.

Left: Detail of small opossum. Courtesy of the National Museum of Natural History.

FISH

We begin with the fish, which are the most abundant of Florissant's vertebrates. They include bowfins, suckers, catfishes, and pirate perches. All of these originally were described by E. D. Cope in the mid-1870s, during the time that he was actively describing new dinosaurs, such as *Camarasaurus,* from much older deposits a few miles south of Florissant. Most of the Florissant fish, except for the pirate perches, were bottom dwellers, and many were tolerant of poor water conditions. Fossil fish seem to be more common in some shale layers than others, perhaps indicating that water conditions in the lake changed through time.

The most primitive fossil fish from Florissant belongs in the order Amiiformes and is represented by *Amia scutata,* a large bowfin (Family Amiidae) that

Figure 212. *(left)* The largest of Florissant's fishes was the bowfin (Order Amiiformes, Family Amiidae), and this specimen of *Amia scutata* measures 43 cm in length. The dorsal fin is long and ribbonlike, giving the bowfin its common name. This fin undulated as the fish remained stationary, and it can be seen extending along more than half of the body length in this fossil. Bowfins belong to a primitive group of fishes that were much more widespread and diverse during the Eocene, but today the only surviving species of bowfin is considered to be a "living fossil." Specimen FMNH-PF-14312, courtesy of the Field Museum of Natural History.

Scale: × 0.65

Figure 213. *(above)* Suckers (Order Cypriniformes, Family Catostomidae) such as this *Amyzon* inhabited the bottom waters of the lake and the tributary streams. The anal fin is closer to the tail in suckers than it is in the other Florissant fish. Specimen UCM-19344, courtesy of the University of Colorado Museum.

Scale: × 0.65

Figure 214. *Ictalurus pectinatus* is a catfish (Order Siluriformes, Family Ictaluridae), and this specimen is preserved to show the bottom side of its broad head and a portion of the vertebral column. Because impressions such as this show the features in reversed relief from the actual fish, it is often easier to study them by making latex peels, which revert the features to normal. Specimen AMNH-FF-8070, courtesy of the Department of Vertebrate Paleontology, American Museum of Natural History.

Scale: × 1.1

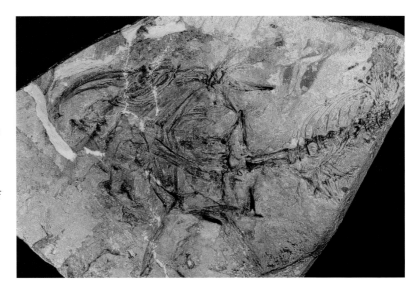

was up to 45 centimeters in length (Figure 212). Only one living species (*Amia calva*) of this family still survives, and it is generally regarded as an example of a "living fossil" because it represents the sole surviving species of a group that was much more diverse and widespread in the geological past. Grande and Bemis recently examined the skeletal morphology and evolution of all of the fossil and living amiid fishes, including 15 specimens from Florissant. Because of its primitive nature, the bowfin is an important link for understanding the evolution of fishes. The only surviving species of the bowfin is restricted to eastern North America but is abundant in freshwater habitats. If the Florissant bowfin was similar to this modern species, it lived in shallow water near the bottom of the lake. The air bladder was able to function as a sort of "lung," which allowed the fish to tolerate stagnant water conditions. It had sharp teeth and

Figure 215. *(left) Trichophanes foliarum* was a pirate perch (Order Percopsiformes, Family Aphredoderidae), which is a family that has only one living species. It was a small fish about 8–10 cm in length and would have made good prey for the large bowfin. Close inspection of this specimen reveals that many of the scales are well preserved. Specimen AMNH-FF-18924A, courtesy of the Department of Vertebrate Paleontology, American Museum of Natural History.

Scale: ×1.7

was an aggressive predator that fed on other fish. The female probably laid eggs in a nest constructed and defended by the male at the bottom of the lake.

The suckers (Order Cypriniformes, Family Catostomidae) are the most diverse group of fish from Florissant and include at least three species in the extinct genus *Amyzon* (Figure 213). These fossil suckers have not been well studied since Cope first described them almost 130 years ago. The suckers were bottom-dwelling fish that sucked the oozes on the floor of Lake Florissant as they fed on algae, organic detritus, and small insect larvae. They were important recyclers of nutrients in the aquatic ecosystem. They may have spawned in the tributary streams around the lake.

The catfishes (Order Siluriformes, Family Ictaluridae) are represented by only two specimens from Florissant (Figure 214), neither of which is complete. The Florissant species of *Ictalurus* belongs to the same genus of catfishes that is predominant in North America today. It was a bottom-feeding scavenger, eating both live and dead organisms including small fish, aquatic insects, snails, and plants. It was probably tolerant of warmer waters and could have lived in the tributary streams as well as in the lake itself. Judging from the modern catfishes, it made nests in the sediment and guarded the eggs and young.

Despite their common name, the pirate perches (Order Percopsiformes, Family Aphredoderidae) are not true perches. The Florissant fossils of this small fish are placed in the extinct genus *Trichophanes* (Figure 215), and several good specimens have been found. The family Aphredoderidae is known from only one living species of the genus *Aphredoderus,* which has many characteristics in common with the fossil genus. *Trichophanes* also shows some affinities with the trout perches (Family Percopsidae), which are known from only two modern species in the genus *Percopsis.* The modern species of *Aphredoderus* and *Percopsis* live in freshwater lakes and ponds and are restricted to North America. *Trichophanes* probably fed on aquatic insects, and as a small fish, it would have been vulnerable prey for the large bowfin.

REPTILES AND AMPHIBIANS: WHERE HAVE ALL THE FOSSILS GONE?

It is a curious thing indeed that Florissant has not produced a single documented discovery of any amphibian or reptile among the more than 40,000 specimens held by museums. Where are the salamanders, frogs, and turtles, all tied to bodies of water for their existence? Were their remains simply not fossilized, or are they so rare in Florissant's fossil record that they have yet to be uncovered? Perhaps they were unable to inhabit the ancient lake because of some

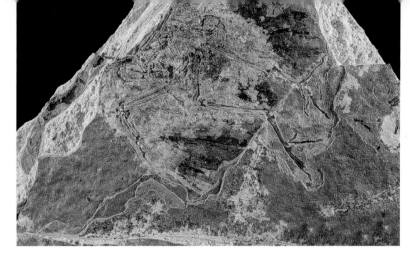

Figure 216. Although the head was not found, this specimen of *Eoculculus cherpinae* (Order Cuculiformes, Family Cuculidae) is excellent for showing feathers between the outstretched wing and leg, possible gut contents in the dark area between the legs, and distinctive characteristics of a cuckoo's foot. It is most similar to modern cuckoos of the Old World. Photograph and description courtesy of Bob Chandler. Specimen DMNH-10682 (holotype), courtesy of the Denver Museum of Nature & Science.

Scale: × 0.6

Figure 217. This complete bird was collected by Samuel Scudder, probably in 1877. It must have attracted great excitement, because it was published just a year later by Allen in 1878. First described as *Paleospiza bella*, a passerine perching bird (Order Passeriformes), it is now considered to be a roller (Order Coraciiformes). The rollers represent a group of arboreal birds that lives today only in the Old World. Specimen MCZ-2222 (holotype), courtesy of the Museum of Comparative Zoology, Harvard University.

Scale: × 0.5

form of toxicity related to the nearby volcanism, yet as we have seen, Lake Florissant *did* contain a variety of aquatic organisms. But the questions remain, and in the meantime, we have nothing to show of amphibians and reptiles.

BIRDS

Fossil birds are rare at Florissant, but three impressive specimens do exist. All come from the lake shale units of the Florissant Formation. Most recently, a new species of cuckoo (Order Cuculiformes, Family Cuculidae), *Eoculculus cherpinae*, was described from a specimen collected in the early 1990s (Figure 216). The head was broken and lost during collection, but the remainder of the skeleton, especially features of the legs and feet, are diagnostic for identifying it as a small arboreal cuckoo. The fossil shows a distinctive cuckoo foot, including the diagnostic character in which two toes point permanently forward, one permanently backward, and one that is reversible forward or backward. *Eoculculus* is related to modern *Cuculus* and is most similar to the common cuckoo and Oriental cuckoo, both of the Old World. These modern cuckoos are typically gray, brown, and white in coloration, often with alternating color stripes on the underside. They occupy a range of habitats including forest, woodland, and scrub, and some are long-distance migrants. Like the living cuckoos, *Eoculculus* probably fed mostly on insects and surely would have had a variety of dietary choices among Florissant's hundreds of insect species.

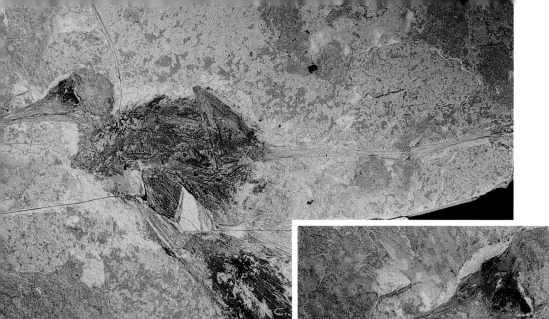
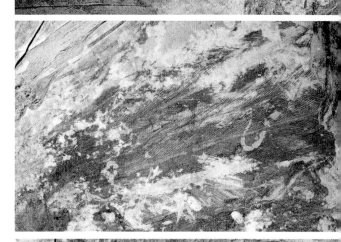
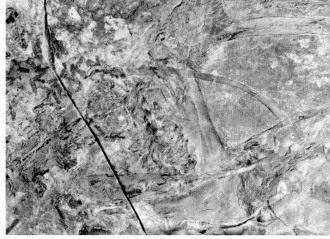

Figure 218. This undescribed shorebird preserves feathers and skeletal features in detail, including the furcula (wishbone) and a bill resembling that of the modern plovers (Order Charadriiformes, Family Charadriidae). The inset photographs show details of the feathers, furcula, and head. Nancy Clare Anderson, who operates the privately owned fossil site, discovered the specimen in 1997. Courtesy of Nancy Clare Anderson, Florissant Fossil Quarry.

Scale: × 0.75
Insets
    Head: × 1.8
    Feathers: × 2.1
    Furcula: × 3.3

Figure 219. This small opossum (Family Didelphidae) is the only known fossil mammal to come from the lake shales. It was an animal that lived in the trees, and this specimen of *Herpetotherium* probably had climbed onto a tree limb overhanging the lake when it fell, drowned, sank, and later was fossilized. The skull, sharp teeth, rib cage, part of one forelimb (the right hand seen near the center of the body), two hind limbs, tail, and the carbonized residue of the hairy carcass are all preserved in this specimen. Specimen USNM-11955, courtesy of the National Museum of Natural History.

Scale: × 1.5

Another nearly complete bird fossil (Figure 217), preserving both the skeletal structure and feathers, is identified as a roller (Order Coraciiformes). In 1878, Allen described it as *Palaeospiza bella* and considered it to be a type of perching passerine (Order Passeriformes). More recent examination, however, has suggested that it is actually a roller, and it may need to be formally reassigned to another genus. In the meantime, we continue to refer to it as *"Palaeospiza" bella*. The modern rollers are restricted to the Old World, including Europe, Africa, southern Asia, and Australia, but they are most diverse in Africa. Like these living relatives, the Florissant rollers probably spent most of their time perched in trees or shrubs but also could walk or hop on the ground. They may have been wide-ranging migrants into far distant regions to the south. They fed on large insects, spiders, or small vertebrates. The modern rollers are mostly bright blue, green, and rusty-brown in coloration. They are named because of their peculiar somersaulting display during courtship flight.

Other birds that appear in the early literature are now considered to be of unknown affinities. One of these was described in 1883 by the paleobotanist Lesquereux, who mistook the true identity of a bird feather and named it *Fontinalis pristina*, a moss! Another bird was described by Cope as *Charadrius sheppardianus*, which he considered to be a plover, although its actual taxonomic affinity needs to be more closely scrutinized. Yet another bird of uncertain classification is *Yalavis tenuipes*, known from a specimen (questionably from Florissant) showing both feet and two feathers.

A new discovery in 1997 uncovered a complete, extremely well preserved specimen of a new species. The long legs and extended beak clearly distinguish it as a shorebird (Figure 218), possibly related to plovers. Although not yet described, it is the finest bird specimen ever to come from the Florissant fossil beds.

Figure 220. Few mammals are preserved at Florissant, and for the most the part, their fossils are only fragmentary. This reconstruction shows the three larger mammals in proportion to their sizes. On the left is the large brontothere, in the middle the *Merycoidodon* (oreodont), and on the right the small *Mesohippus* (horse).

New discoveries such as this assure that the study of Florissant's paleontology remains an exciting story, and one that is never completely concluded.

MAMMALS: THE SMALL PIECES OF THE LARGEST ANIMALS

When you go into a forest today, you always see the plants and usually many insects as well. Some of the common birds are easy to spot, but seeing one of the larger mammals is often an event that might cause you to take notice and pause for a moment. And so it is with the fossil mammals from Florissant. They are rare, and you only get to see a glimpse of them. But even still, you know that there must have been many more that lived in this ancient community.

The only mammal fossil recorded from the lake shales consists of the entire body of a small opossum (Family Didelphidae) belonging in the extinct genus *Herpetotherium* (originally placed into *Peratherium* cf. *P. hunteri*) (Figure 219). It was a small animal related to the modern mouse opossums that live today from Mexico to South America. *Herpetotherium* was probably active mainly during the evening and night, and was a forest dweller that lived in trees or on the ground. Its hands and feet, and probably also a prehensile tail, made it well adapted to an arboreal habitat in the branches of trees. It was omnivorous, feeding on insects, other small vertebrates, and fruit. Opossums later became extinct in North America during the Miocene Epoch, but they survived in South America and later returned to North America when the Central American land bridge formed. The modern representatives of the family Didelphidae live only in the New World.

Figure 221. The primitive *Mesohippus* horse (Family Equidae) is represented by this well-preserved lower mandible, illustrated in both occlusal and lateral views. This horse was about the size of a collie, and the mandible measures about 12 cm in length. The front incisors were used to nibble the foliage and leaves from plants, and the premolars and molars had folded, W-shaped ridges that were an adaptation for grinding this fodder. *Mesohippus* had three-toed feet and lived in both dense forests and open woodlands. Specimen UCM-65951, from the collection of Florissant Fossil Beds National Monument and housed at the University of Colorado Museum.

Scale:
top: × 0.75
bottom: × 1.1

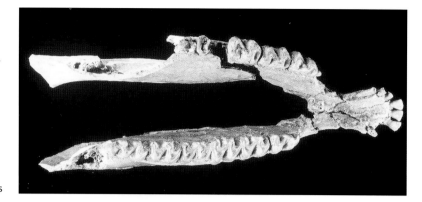

The lower mudstone unit of the Florissant Formation have produced fragmentary evidence of a horse, a brontothere, and an oreodont (Figure 220). The horse (Family Equidae) was a small, three-toed form belonging to the genus *Mesohippus*. It was about the size of a collie and is known at Florissant from a complete lower jaw (Figure 221). The teeth were low and had chewing surfaces that were folded into W-shaped ridges, which was an adaptation for a browsing diet on the foliage of herbs as well as the leaves of shrubs or small trees. The agile, three-toed foot allowed these small horses to move with ease through the shrubbery of the forest understory. Their habitat ranged from the dense lakeside forest to the open woodlands.

The largest mammal known to have lived in the ancient Florissant community was the brontothere (Family Brontotheriidae). It was a large, heavy animal—almost elephant-sized and more than two meters in height at the shoulder—and had a pair of thick horns (Figure 220). It was a herbivore that browsed on soft vegetation, using its small incisors to strip the softer foliage from herbs, shrubs, and trees and chewing with its massive, low-crowned teeth. The tremendous size of the brontothere would have restricted its movements, particularly in areas of the valley bottom near the lake where the vegetation was dense, and it probably was confined primarily to the more open clearings and woodlands. Brontotheres are an extinct group, and they are one of the few Florissant fossils belonging to an extinct family. But the fossils that Florissant has produced of these giant beasts—fragmentary remains of an atlas vertebra and a scattering of small pieces of tooth enamel—are not impressive. Still, the presence of brontotheres adds a significant perspective to Florissant's whole story, not only about what lived in the ancient community, but also about extinction and how it relates to Florissant's age. Brontotheres lived only during the Eocene Epoch, first appearing as relatively small horselike animals and eventually evolving into huge rhinoceros-like creatures. At the time the Eocene ended—only a few hundred thousand years after Florissant's fossils were laid to rest—the fossil record throughout the Northern Hemisphere documents the brontothere's extinction. This widespread and well-documented extinction event at the close of the Eocene makes brontothere fossils especially useful as time indicators, and the presence of a brontothere at Florissant is one of the strongest lines of evidence that Florissant is Eocene, and not Oligocene, in age. Brontothere fossils are much more common in late Eocene deposits from the Badlands of South Dakota and the Great Plains of northeastern Colorado, and it may be that brontotheres only occasionally wandered into the higher elevations around Florissant.

Another herbivorous mammal was the oreodont (Family Merycoidodontidae), documented from one jaw fragment of which only a plaster cast remains. Like the brontotheres, this group represents an extinct family, although oreodonts persisted for many millions of years after the brontotheres and were common during the Oligocene and Miocene Epochs. They are known only from North and Central America. Because there are no living members of this family, oreodonts are difficult to compare with modern mammals. They usually are reconstructed as sheeplike or piglike in general appearance, yet they were quite different. Oreodonts had protruding upper canines, and the molar teeth had a series of ridged, V-shaped cusps that were adapted for browsing. Oreodonts occupied the habitats of forests, woodlands, and clearings.

RELATED REFERENCES

Florissant fossil fish: Cope (1884); Grande and Bemis (1998); Lundberg (1975); Rosen and Patterson (1969)
Florissant fossil birds: Allen (1878a, 1878b); Chandler (1999)
Florissant fossil mammals: Gazin (1935)
Taxonomy and ecology of related modern vertebrates: Banister and Campbell (1985); Cameron and Harrison (1978); Migdalski and Fichter (1983); Perrins and Middleton (1985)

EPILOGUE

Florissant's Changing Perspective in Time

In an interview during his last visit to Florissant in 1979, Harry D. MacGinitie summarized well the perspective of time at Florissant with the following remark: "We get all tangled up with the present. The present is just a little flick in time between the past and the future. Things keep going on and on. We are just in this particular little time interval, and it seems so important to us."

What do we really learn by studying ancient worlds like Florissant? Throughout the chapters of this book, we have looked at a few of the geologic events that created a bygone landscape, we have seen how climate and communities have changed over time, and we have been introduced to a tremendously large cast of characters among the world of extinct plants, insects, and vertebrates. But in the end, this entire book is about our human perspective of this ancient world, revealed by what paleontologists have uncovered over the relatively short time of 130 years. These paleontologists have used the rigorous methods of scientific inquiry to create the body of knowledge that makes possible our understanding of the fossils of Florissant. They have done this by carefully collecting this record of ancient life, by analyzing the context of the rocks in which the fossils occur, by making comparisons with living organisms and communities, and by naming hundreds upon hundreds of new species. But these taxonomic names would have had no meaning whatsoever to the ancient organisms that once lived at Florissant, for theirs was simply a world of survival, not the quest for understanding that human intelligence has since evolved to seek.

To fully appreciate the broad perspective of geologic time at Florissant, we need to pause, in conclusion, to consider what's really happening at Florissant. One hundred thirty years is not very

Left: Detail of the elderberry, *Sambucus newtoni.* Courtesy of the National Museum of Natural History.

long to study and impart meaning to fossils that are more than 34 *million* years old. What *has* happened in that much longer span of geologic time? The warm temperate climate long ago cooled, shortly after Florissant's fossils were deposited. The species that lived here in the late Eocene have become extinct, and many of their surviving relatives now live as far away as China and Africa. An ancient biotic community has disappeared, making way for the *new* communities of plants and animals that inhabit Florissant today. Slowly, over the past few million years, erosion has exposed some of Florissant's ancient rocks and fossils for paleontologists to collect, but it also has carried much of the sediment of these rocks and fossils to the Mississippi River delta. And what does this tell us of the future here? We speak of preservation of the fossil beds, a noble objective achievable in our own time, but ultimately, one that will not, and cannot, last in perpetuity. Surely much of what remains at Florissant also is destined to find its way down the Mississippi River eventually. Nature doesn't remove these fossils in the same way that paleontologists do.

Humans also are concerned about the potential impacts of changing global climate. The fossil record provides the hindsight for predicting future biotic responses to climate change, and for understanding the potential impacts of human activities on the environment for other species, yet sites such as Florissant also reveal that climate change brings not only extinction and the loss of what was, but also new opportunities for life to go forth with new innovations into the environments of the future. The study of geology and paleontology at places like Florissant clearly shows that the world is, if nothing else, an ever-changing, evolving place. But in the meantime, we view it from the little flick in time that seems so important to us, just as *Prodryas persephone* did in its world 34 million years ago.

APPENDIX 1

A Complete Listing of the Fossil Organisms from Florissant

This appendix lists the most recently updated taxonomic names for all of the described fossil organisms from the Florissant Formation. These taxonomic names reflect any changes that have been proposed in the literature since the species were originally described, although the information presented here is continually being revised and updated in the Florissant database website. Originally described species that were later placed into synonymy with other species are not listed here. In addition, some of the generic names, especially for extant genera of insects, are revised according to the current names that are used for these genera; summaries of such revisions are found in contemporary sources such as *Nomina Insecta Nearctica* (Poole and Gentili 1996–1997; also available as a website). That is, fossils assigned decades ago by workers such as Scudder to extant generic names that are no longer in use are shown in this appendix according to the more recent generic synonymies that are based upon extant species. These cases are indicated by listing first the generic name that is currently recognized, followed in brackets by the name that was last published for this taxon at Florissant. In such cases, the fossils themselves have not been reevaluated to determine with certainty whether or not they conform to those generic changes that are based on the extant species of those genera. Most names of extinct genera are retained as they were originally proposed and are indicated here as extinct genera with the symbol †. This symbol also is used to denote fossil form genera (i.e., those that lack diagnostic characteristics by which they could be assigned confidently to particular extant genera). Generic names of Coleoptera within quotation marks denote assignments according to Carpenter (1992). The following lists are intended to provide a synthesis of these taxonomic data, but any implied changes to the most recently published names are not to be considered as formal taxonomic revisions proposed by this work. Taxonomic names are continually subject to revision, and although the ones given in this appendix are current in the literature, it should be emphasized that many taxa are in need of closer scrutiny, especially the insects.

Higher taxonomic ranks such as family have been updated to reflect recent concepts of taxonomy. In general, these follow Takhtajan (1997) for the plants and Borror et al. (1989) for the insects. Although these references provide a consistent source for higher taxonomic classification, such classifications are always subject to revision, and it should be noted that other recent monographs may have proposed classifications that differ from those used here. Separate fossil plant lists are given for (1) wood, (2) leaves, fruits, and seeds, and (3) pollen and spores (two lists, one based on palynomorphic names and the other based on taxonomic names). The updates to the taxonomic list of fossil insects were compiled by Boyce Drummond as a research project for the National Park Service.

The capitalized abbreviation following a taxonomic name indicates the museum at which the primary type specimen is held (only if the type specimen is from Florissant), and these abbreviations are defined in Appendix 2. The primary type specimens include holotypes, syntypes, and lectotypes. In cases where more than one museum is indicated for the same taxon, either the holotype specimen has its corresponding halves (part and counterpart) at different museums or the taxon is based on syntypes that are at different museums. Type specimens that are described in publication but that have not been located in museum collections are indicated as such. In rare cases, specimens that were indicated as types in museum collections could not be correlated with a publication documenting their valid description, and these are included in this appendix with an appropriate notation. Instances in which the actual type specimen is obscured by nomenclatural problems are denoted with "[?]." Taxa not named to species are followed by an acronym to show the location of the voucher specimen.

Many of the Florissant specimens have been dealt with in multiple publications and under various taxonomic combinations. Readers who would like to obtain more specific information than that provided by this abridged appendix are invited to use the Florissant database website (accessible by link from the website for Florissant Fossil Beds National Monument at www.nps.gov/flfo). This database includes the entire taxonomic and publication history for more than 99 percent of all Florissant fossils that have ever been referenced in publication, as well as all of the museum catalog data that pertain to the type specimens.

PLANTS

WOODS

Division Coniferophyta

ORDER CONIFERALES (conifers)

Family TAXODIACEAE [~CUPRESSACEAE] (bald cypress family)

†*Sequoioxylon pearsallii* Andrews 1936 [also referred to *Sequoia affinis* Lesquereux 1876, in MacGinitie 1953]: FLFO [type is the Big Stump]

Division Magnoliophyta (flowering plants)

Class Magnoliopsida (dicotyledonous flowering plants)

ORDER URTICALES

Family ULMACEAE (elm family)

cf. *Zelkova* sp. Wheeler 2001: DMNH

†*Zelkovoxylon chadronensis* Wheeler 2001: DMNH

ORDER FABALES

Family FABACEAE [= LEGUMINOSAE] (legume family)

cf. *Robinia zirkelii* (Platen) Matten, Gastaldo, & Lee 1977 [of Wheeler 2001]: [type not from Florissant]

ORDER SAPINDALES

Family SAPINDACEAE (soapberry family)

cf. *Koelreuteria* sp. Wheeler 2001: DMNH

INCERTAE SEDIS (unknown affinity)

†*Chadronoxylon florissantensis* Wheeler 2001: DMNH

LEAVES, FRUITS, SEEDS, AND FLOWERS

Division Bryophyta (mosses)

Family GRIMMIACEAE

†*Plagiopodopsis cockerelliae* (Britton & Hollick) Steere 1947: YPM

†*Plagiopodopsis scudderi* Britton & Hollick 1915: USNM

Division Sphenophyta

ORDER EQUISETALES (horsetails)

Family EQUISETACEAE

Equisetum florissantense Cockerell 1915: UCM

Division Pteridophyta

ORDER FILICALES (ferns)

Family ASPLENIACEAE

Dryopteris guyottii (Lesquereux) MacGinitie 1953: USNM

Division Gnetophyta

ORDER GNETALES

Family EPHEDRACEAE (joint-fir family)

Ephedra miocenica Wodehouse 1934: YPM

Division Coniferophyta

ORDER CONIFERALES (conifers)

Family TAXACEAE (yew family)

Torreya geometrorum (Cockerell) MacGinitie 1953: UCM

Family CUPRESSACEAE (cypress family)

Chamaecyparis linguaefolia (Lesquereux) MacGinitie 1953: USNM

Family TAXODIACEAE [~CUPRESSACEAE] (bald cypress family)

Sequoia affinis Lesquereux 1876: USNM

Family PINACEAE (pine family)

Abies rigida Knowlton 1923 [of Wolfe and Schorn 1990]: [type not from Florissant]

Picea magna MacGinitie 1953: [type not from Florissant]

Pinus florissantii Lesquereux 1883 (subgenus Haploxylon, cone): USNM

Pinus macginitiei Axelrod 1986 (subgenus Diploxylon, two- to three-needled foliage and seeds): UCMP

Pinus wheeleri Cockerell 1908 (subgenus Haploxylon, five-needled foliage): USNM

Pinus sp. (subgenus Diploxylon, five-needled foliage) [= *Pinus wheeleri* Cockerell of MacGinitie 1953, pl. 18, fig. 11 only]: UCMP [not a type specimen]

Pinus sp. (subgenus Diploxylon, cone) [= *P. wheeleri* Cockerell 1908, cone only): USNM

Pinus sp. (subgenus Haploxylon) [= *Pinus* leaves of MacGinitie 1953, pl. 18, fig. 3]: UCMP [not a type specimen]

Division Magnoliophyta (flowering plants)

Class Magnoliopsida (dicotyledonous flowering plants)

ORDER ARISTOLOCHIALES

Family ARISTOLOCHIACEAE (birthwort family)

Aristolochia mortua Cockerell 1908: USNM

ORDER LAURALES

Family LAURACEAE (laurel family)

Lindera coloradica MacGinitie 1953: UCMP

Persea florissantia MacGinitie 1953: DMNH

ORDER BERBERIDALES

Family BERBERIDACEAE (barberry family)

Mahonia marginata (Lesquereux) Arnold 1936: USNM

Mahonia obliqua MacGinitie 1953: UCMP

Mahonia subdenticulata (Lesquereux) MacGinitie 1953: USNM

ORDER HAMAMELIDALES

Family PLATANACEAE (sycamore family)

Platanus florissantii MacGinitie 1953: DMNH

ORDER FAGALES

Family FAGACEAE (beech family)

Castanea dolichophylla Cockerell 1908: UCM, USNM

†*Fagopsis longifolia* (Lesquereux) Hollick 1909: USNM

Quercus dumosoides MacGinitie 1953: USNM

Quercus knowltoniana Cockerell 1908: YPM

Quercus lyratiformis Cockerell 1908: USNM

Quercus mohavensis Axelrod 1939 [of MacGinitie 1953]

Quercus orbata MacGinitie 1953: UCMP

Quercus peritula Cockerell 1908: USNM

Quercus predayana MacGinitie 1953: USNM

Quercus scottii (Lesquereux) MacGinitie 1953: USNM

Quercus scudderi Knowlton 1916: USNM

ORDER CORYLALES

Family BETULACEAE (birch family)

†*Asterocarpinus perplexans* (Cockerell) Manchester & Crane 1987: USNM

†*Paracarpinus fraterna* (Lesquereux) Manchester & Crane 1987: USNM

ORDER JUGLANDALES

Family JUGLANDACEAE (walnut family)

Carya florissantensis Manchester 1987: FMNH

Carya libbeyi (Lesquereux) MacGinitie 1953: USNM, YPM

Juglans(?) *sepultus* Cockerell 1908: USNM

ORDER STYRACALES

Family STYRACACEAE (storax family)

Halesia reticulata MacGinitie 1953: UCMP

ORDER SALICALES

Family SALICACEAE (willow family)

Populus crassa (Lesquereux) Cockerell 1908: USNM

Salix coloradica MacGinitie 1953: UCMP

Salix libbeyi Lesquereux 1883: USNM

Salix ramaleyi Cockerell 1906: UCM

Salix taxifolioides MacGinitie 1953: USNM

ORDER MALVALES

Family TILIACEAE (basswood family)

Tilia populifolia Lesquereux 1883: USNM

Family STERCULIACEAE (cocoa family)

†*Florissantia speirii* (Lesquereux) Manchester 1992: USNM

ORDER URTICALES

Family ULMACEAE (elm family)

†*Cedrelospermum lineatum* (Lesquereux) Manchester 1989: USNM

Celtis mccoshii Lesquereux 1883: USNM

Ulmus tenuinervis Lesquereux 1874: USNM

Family MORACEAE (mulberry family)

Morus symmetrica Cockerell 1908: USNM

Family CANNABACEAE (marijuana family)

Humulus florissantella (Cockerell) MacGinitie 1969: USNM

ORDER EUPHORBIALES

Family EUPHORBIACEAE (spurge family)

Euphorbia minuta (Knowlton) MacGinitie 1953: USNM

ORDER THYMELAEALES

Family THYMELAEACEAE

Daphne septentrionalis (Lesquereux) MacGinitie 1953: USNM

ORDER SAXIFRAGALES

Family GROSSULARIACEAE (currant family)

Ribes errans MacGinitie 1953: UCMP

ORDER ROSALES

Family ROSACEAE (rose family)

Amelanchier scudderi Cockerell 1906: UCM

Cercocarpus myricaefolius (Lesquereux) MacGinitie 1953: USNM

Crataegus copeana (Lesquereux) MacGinitie 1953: USNM

Crataegus hendersonii (Cockerell) MacGinitie 1953: UCM

Crataegus nupta (Cockerell) MacGinitie 1953: UCM

Holodiscus lisii Schorn 1998: UCMP

Malus florissantensis (Cockerell) MacGinitie 1953: USNM

Malus pseudocredneria (Cockerell) MacGinitie 1953: USNM

Prunus gracilis (Lesquereux) MacGinitie 1953: USNM

Rosa hilliae Lesquereux 1883: USNM

Rubus coloradense (MacGinitie) Wolfe & Tanai 1987: UCMP

Vauquelinia coloradensis (Knowlton) MacGinitie 1953: USNM

Vauquelinia liniara MacGinitie 1953: UCMP

ORDER MYRTALES

Family ONAGRACEAE (evening primrose family)

Onagraceous flower [Kirchner 1898]: USNM

Family MYRTACEAE (myrtle family)

"*Eugenia*" *arenaceaeformis* (Cockerell) MacGinitie 1953: USNM

ORDER FABALES

Family FABACEAE [= LEGUMINOSAE] (legume family)

Astragalus wilmattae Cockerell 1908: NHM, YPM

†*Caesalpinites acuminatus* (Lesquereux) MacGinitie 1953: USNM

†*Caesalpinites coloradicus* MacGinitie 1953: USNM

Cercis parvifolia Lesquereux 1883: USNM

Conzattia coriacea MacGinitie 1953: UCMP

†*Leguminosites lespedezoides* MacGinitie 1953: UCMP

†*Phaseolites dedal* MacGinitie 1953: UCMP

Prosopis linearifolia (Lesquereux) MacGinitie 1953: [specimen has not been located]

Robinia lesquereuxi (Ettingshausen) MacGinitie 1953: [?]

Vicia sp. Knowlton 1916: USNM

ORDER SAPINDALES

Family STAPHYLEACEAE (bladdernut family)

Staphylea acuminata Lesquereux 1874: USNM

Family SAPINDACEAE (soapberry family)

Athyana haydenii (Lesquereux) MacGinitie 1953: [?]

Cardiospermum terminale (Lesquereux) MacGinitie 1953: USNM

Dodonaea umbrina MacGinitie 1953: UCMP

Koelreuteria allenii (Lesquereux) Edwards 1927: USNM

Sapindus coloradensis Cockerell 1908: UCM, USNM

Thouinia straciata MacGinitie 1953: UCM

Family ACERACEAE [=SAPINDACEAE] (maple family)

Acer florissantii Kirchner 1898: USNM

Acer macginitiei Wolfe & Tanai 1987: UCMP

Dipteronia brownii McClain & Manchester 2001: [type not from Florissant]

†*Bohlenia insignis* (*Lesquereux*) Wolfe & Wehr 1987: USNM

ORDER RUTALES

Family RUTACEAE (citrus family)

Ptelea cassiodes (Lesquereux) MacGinitie 1953: USNM

Ptelea-like fruit [Manchester 2001] [= *Brachyruscus alleni* (Cockerell)]: UCM, NHM

Family SIMAROUBACEAE (tree of heaven family)

Ailanthus americana Cockerell 1908: UCM

Family uncertain, aff. SIMAROUBACEAE

†*Chaneya tenuis* (Lesquereux) Wang & Manchester 2000: USNM

Family MELIACEAE (mahogany family)

Cedrela lancifolia (Lesquereux) Brown 1937: USNM

Trichilia florissantii (Lesquereux) MacGinitie 1953: USNM

ORDER BURSERALES

Family BURSERACEAE (torchwood family)

Bursera serrulata (Lesquereux) MacGinitie 1953: USNM

Family ANACARDIACEAE (cashew family)

Astronium truncatum (Lesquereux) MacGinitie 1953: USNM

Cotinus fraterna (Lesquereux) Cockerell 1906: USNM

Rhus lesquereuxi Knowlton & Cockerell 1919 [of MacGinitie 1953]: [?]

Rhus obscura (Lesquereux) MacGinitie 1953: USNM

Rhus stellariaefolia (Lesquereux) MacGinitie 1953: USNM

Rhus vexans [= *Schmaltzia vexans* (Lesquereux) MacGinitie 1953]: USNM

ORDER CELASTRALES

Family CELASTRACEAE (spindle tree family)

Celastrus typica (Lesquereux) MacGinitie 1953: USNM

ORDER RHAMNALES

Family RHAMNACEAE (buckthorn family)

Colubrina spireaefolia (Lesquereux) MacGinitie 1953: USNM

†*Rhamnites pseudo-stenophyllus* (Lesquereux) MacGinitie 1953: USNM

Ziziphus florissantii (Lesquereux) MacGinitie 1953: USNM

ORDER VITALES

Family VITACEAE (grape family)

Parthenocissus osbornii (Lesquereux) MacGinitie 1953: USNM

Vitis sp. [Manchester 2001] [= *Vitis florissantella* Cockerell, specimen of MacGinitie 1953, plate 67 only]: UCM [not a type specimen]

ORDER HYDRANGEALES

Family HYDRANGEACEAE (hydrangea family)

Hydrangea fraxinifolia (Lesquereux) Brown 1937: USNM

Philadelphus parvulus Becker 1961: [?]

ORDER EUCOMMIALES

Family EUCOMMIACEAE (hard-rubber tree family)

Eucommia sp. Manchester 2001: USNM

ORDER ARALIALES

Family ARALIACEAE (ginseng family)

Oreopanax dissecta (Lesquereux) MacGinitie 1953: USNM

Araliaceaous infructescence Manchester 2001: UF, WC

ORDER ADOXALES

Family SAMBUCACEAE (elder family)

Sambucus newtoni Cockerell 1908: NHM, UCM

ORDER DIPSACALES

Family CAPRIFOLIACEAE (honeysuckle family)

†*Diplodipelta reniptera* (Becker) Manchester & Donoghue 1995: [type is not from Florissant]

ORDER APOCYNALES

Family APOCYNACEAE (sweetsop family)

†*Apocynospermum* sp. Manchester 2001: UCMP

ORDER CONVOLVULALES

Family CONVOLVULACEAE (morning-glory family)

†*Convolvulites orchitus* MacGinitie 1953: UCMP

ORDER OLEALES

Family OLEACEAE (olive family)

Osmanthus praemissa (Lesquereux) Cockerell 1906: USNM

Class Liliopsida (monocotyledonous flowering plants)

ORDER SMILACALES

Family SMILACACEAE (greenbrier family)

Smilax labidurommae Cockerell 1914: UCM

ORDER DIOSCOREALES

Family DIOSCOREACEAE (yam family)

Dioscorea sp. Manchester 2001: UCMP

ORDER CYPERALES

Family CYPERACEAE (sedge family)

†*Cyperacites lacustris* MacGinitie 1953: MPM

ORDER POALES

Family POACEAE [= GRAMINEAE] (grass family)

Stipa florissantii (Knowlton) MacGinitie 1953: USNM

ORDER ARECALES

Family ARECACEAE [= PALMAE] (palm family)

†*Palmites* sp. Manchester 2001: UWBM

ORDER POTAMOGETONALES

Family POTAMOGETONACEAE

Potamogeton geniculatus Al Br. [of Manchester 2001]: [type is not from Florissant]

ORDER ARALES

Family LEMNACEAE

†*Limnobiophyllum scutatum* (Dawson) Krassilov 1995 [of Manchester 2001]: [type is not from Florissant]

ORDER TYPHALES

Family TYPHACEAE (cattail family)

Typha lesquereuxi Cockerell 1906: [?]

Incertae sedis (classification uncertain and/or genera/form genera in need of revision)

†*Antholithes amoenus* Lesquereux 1883: USNM

†*Antholithes pedilioides* Cockerell 1915: UCM

†*Archaeomnium brownii* (Kirchner) LaMotte 1952: USNM

†*Carpites gemmaceus* Lesquereux 1883: USNM

†*Carpites miliodes* Lesquereux 1883: USNM

Croton(?) *furculatum* Cockerell 1909: UCM, NHM

†*Deviacer* sp. Manchester 2001: USNM

Ficus bruesi Cockerell 1910: UCM, NHM

Ficus florissantella Cockerell 1908: USNM

†*Hypnum haydenii* Lesquereux 1876: USNM

Ilex knightiafolia Lesquereux 1883: USNM

†*Jungermanniopsis cockerellii* Howe & Hollick 1922: YPM

Lomatia lineata (Lesquereux) MacGinitie 1953 (leaves only)

†*Lomatites spinosa* (Lesquereux) Cockerell 1908: USNM

†*Najadopsis rugulosa* Lesquereux 1883: USNM

†*Palaeopotamogeton florissantii* Knowlton 1916: USNM

Panax andrewsii Cockerell 1908: UCM, NHM

Pellea antiquella Cockerell 1908: USNM

†*Phenanthera petalifera* Hollick 1907: YPM

†*Populites heeri* (Lesquereux) MacGinitie 1953: USNM

Populus pyrifolia Kirchner 1898: USNM

Potamogeton(?) *verticillatus* Lesquereux 1883: USNM

Quercus balaninorum Cockerell 1908: USNM

Sterculia rigida Lesquereux 1883: USNM

POLLEN AND SPORES (OF WINGATE AND NICHOLS 2001)

Algal and Probable Algal Microfossils

Botryococcus sp.

Pediastrum sp.

Ovoidites elongates (Hunger) Krutzsch 1959 [of Wingate and Nichols 2001]

Ovoidites ligneolus Potonié ex Krutzsch 1959 [of Wingate and Nichols 2001]

Ovoidites sp.

Algal spores, genus and species indeterminate

Catinipollis geiseltalensis Krutzsch 1966 [of Wingate and Nichols 2001]

Pteridophyte Spores: Monolete

Laevigatosporites ovatus Wilson & Webster 1946 [of Wingate and Nichols 2001]

Polypodiisporonites afavus (Krutzsch) Frederiksen 1980 [of Wingate and Nichols 2001]

Polypodiidites sp. cf. *P. secundus* (Potonié) Krutzsch 1963 [of Wingate and Nichols 2001]

Bryophyte and Pteridophyte Spores: Trilete

Lygodiumsporites adriennis (Potonié & Gelletich) Potonié, Thomson & Thiergart ex Potonié 1956 [of Wingate and Nichols 2001]

Deltoidospora sp.

Baculatisporites sp. A

Baculatisporites sp. B

Biretisporites sp.

Polycingulatisporites sp.

Lusatisporis sp. cf. *L. perinatus* Krutzsch, Sontag & Pacltová 1963 [of Wingate and Nichols 2001]

Heliosporites sp.

Foveotriletes sp.

Stereisporites sp.

Echinatisporis sp.

Reticulate trilete, genus and species indeterminate

Rugulate trilete, genus and species indeterminate

Spore Massulae

Azolla cretacea Stanley 1965 [of Wingate and Nichols 2001]

Gymnosperm Pollen: Bisaccate

?*Abiespollenites* sp.

Piceapollis sp.

Pristinuspollenites sp. cf. *P. microsaccus* (Couper) Tschudy 1973 [of Wingate and Nichols 2001]

Pityosporites sp. A

Pityosporites sp. B

Podocarpidites sp.

Gymnosperm Pollen: Monosaccate

Zonalapollenites sp.

Gymnosperm Pollen: Inaperturate

Taxodiaceaepollenites hiatus (Potonié) Kremp ex Potonié 1958 [of Wingate and Nichols 2001]

cf. *Taxodiaceaepollenites* sp. A

Taxodiacites sp. cf. *T. verrucosus* Botscharnikova 1960 [of Wingate and Nichols 2001]

Sequoiapollenites sp. cf. *S. rotundus* Krutzsch 1971 [of Wingate and Nichols 2001]

Gymnosperm Pollen: Polyplicate

Ephedripites claricristatus (Shakhmundes) Krutzsch 1970 [of Wingate and Nichols 2001]

Ephedripites exiguus (Frederiksen) Wingate & Nichols 2001

Angiosperm Pollen: Monosulcate

Arecipites sp. cf. *A. pertusus* (Elsik) Nichols, Ames, & Traverse 1973 [of Wingate and Nichols 2001]

Liliacidites sp.

?*Nupharipollenites* sp.

Angiosperm Pollen: Tricolpate, Sculpture Various

    Cupuliferoidaepollenites liblarensis (Thomson in Potonié, Thomson, & Thiergart) Potonié 1960 [of Wingate and Nichols 2001]

    Fraxinoipollenites medius Frederiksen 1973 [of Wingate and Nichols 2001]

    Salixipollenites sp. A

    Salixipollenites sp. B

    Salixipollenites sp. C

    Rousea araneosa (Frederiksen) Frederiksen 1980 [of Wingate and Nichols 2001]

    Rousea sp.

    Quercoidites microhenricii (Potonié) Potonié 1960 [of Wingate and Nichols 2001]

    Cercidiphyllites sp.

    Aceripollenites striatus (Pflug) Thiele-Pfeiffer 1980 [of Wingate and Nichols 2001]

    Aceripollenites sp.

    Tricolpate sp. A

    Tricolpate sp. B

    Tricolpate sp. C

    Tricolpate sp. D

Angiosperm Pollen: Tricolporate, Psilate to Scabrate

    Tricolpopollenites parmularius (Potonié) Thomson & Pflug 1953 [of Wingate and Nichols 2001]

    Siltaria sp. cf. *S. scabriextima* Traverse 1955 [of Wingate and Nichols 2001]

    Siltaria sp. cf. *S. pacata* (Pflug in Thomson & Pflug) Frederiksen 1980 [of Wingate and Nichols 2001]

    Rhamnacidites delicatus Frederiksen in Frederiksen et al. 1983 [of Wingate and Nichols 2001]

Angiosperm Pollen: Tricolporate, Reticulate

    Araliaceoipollenites profundus Frederiksen 1980 [of Wingate and Nichols 2001]

    Araliaceoipollenites euphorii (Potonié) Potonié 1960 [of Wingate and Nichols 2001]

    cf. *Horniella brevicolpata* Frederiksen et al. 1983 [of Wingate and Nichols 2001]

    Horniella sp. A

    Horniella sp. B

    Margocolporites sp. cf. *M. vanwijhei* Germeraad, Hopping, & Muller 1968 [of Wingate and Nichols 2001]

    Rhoipites sp. A

    Rhoipites sp. B

    Rhoipites sp. C

    Rhoipites sp. D

    Rhoipites sp. E

    Rhoipites sp. F

    Tricolporate sp. A

    Tricolporate sp. B

    Tricolporate sp. C

Angiosperm Pollen: Tricolporate, Striate

    Ailanthipites berryi Wodehouse 1933 [of Wingate and Nichols 2001]

    Ailanthipites sp. A

    Ailanthipites sp. B

    Alangiopollis sp.

    Tricolporate sp. D

    Tricolporate sp. E

Angiosperm Pollen: Tricolporate, Verrucate

    ?*Quercipollenites* sp. cf. pollen of *Fagopsis longifolia* (Lesquereux) Hollick [of Wingate and Nichols 2001]

    Verrutricolporites sp. A

    Verrutricolporites sp. B

    Slowakipollis hippophaëoides Krutzsch 1962 [of Wingate and Nichols 2001]

Angiosperm Pollen: Tricolporate, Echinate

    Asteraceae sp. A

Angiosperm Pollen: Tetracolporate

    Tetracolporopollenites brevis Frederiksen 1980 [of Wingate and Nichols 2001]

    Tetracolporopollenites sp.

    ?*Meliapollis* sp.

Angiosperm Pollen: Syncolporate

    Boehlensipollis sp. cf. *B. granulata* Frederiksen et al. 1983 [of Wingate and Nichols 2001]

    Myrtaceidites parvus Cookson & Pike 1954 [of Wingate and Nichols 2001]

Cupanieidites sp. A

Cupanieidites sp. B

Angiosperm Pollen: Tribrevicolporate

> *Bombacacidites* sp. cf. *B. reticulatus* Krutzsch 1961 sensu Frederiksen 1988 [of Wingate and Nichols 2001]
>
> *Bombacacidites* sp. cf. *B. nanobrochatus* Frederiksen et al. 1983 sensu Frederiksen 1988 [of Wingate and Nichols 2001]
>
> *Bombacacidites* sp. A
>
> *Bombacacidites* sp. B
>
> *Lonicerapollis* sp. A
>
> *Lonicerapollis* sp. B
>
> *Lonicerapollis* sp. C
>
> *Intratriporopollenites* sp. cf. *I. instructus* (Potonié) Thomson & Pflug 1953 [of Wingate and Nichols 2001]

Angiosperm Pollen: Triporate

> *Caryapollenites veripites* (Wilson & Webster) Nichols & Ott 1978 [of Wingate and Nichols 2001]
>
> *Corsinipollenites oculus-noctis parvus* (Doktorowicz-Hrebnicka) Krutzsch 1968 [of Wingate and Nichols 2001]
>
> *Corsinipollenites parviangulus* Frederiksen et al. 1983 [of Wingate and Nichols 2001]
>
> *Cricotriporites intrastructurus* (Krutzsch & Vanhoorne) Wingate & Nichols 2001
>
> ?*Cricotriporites* sp.
>
> *Momipites coryloides* Wodehouse 1933 [of Wingate and Nichols 2001]
>
> *Momipites microfoveolatus* (Stanley) Nichols 1973 [of Wingate and Nichols 2001]
>
> *Momipites triradiatus* Nichols 1973 [of Wingate and Nichols 2001]
>
> *Momipites ventifluminis* Nichols & Ott 1978 [of Wingate and Nichols 2001]
>
> *Triatriopollenites subtriangulus* (Stanley) Frederiksen 1979 [of Wingate and Nichols 2001]
>
> *Nudopollis* sp.
>
> *Trivestibulapollenites betuloides* Pflug in Thomson & Pflug 1953 [of Wingate and Nichols 2001]

Triporate sp. A

Triporate sp. B

Triporate sp. C

Angiosperm Pollen: Zonoporate

> *Alnipollenites verus* Potonié ex Potonié 1931 [of Wingate and Nichols 2001]
>
> *Reevesiapollis triangulus* (Mamczar) Krutzsch 1970 [of Wingate and Nichols 2001]
>
> ?*Reevesiapollis* sp.
>
> *Ulmipollenites undulosus* Wolff 1934 [of Wingate & Nichols 2001]
>
> Zonoporate sp. A
>
> Zonoporate sp. B
>
> Zonoporate sp. C

Angiosperm Pollen: Pantoporate, Psilate

> *Juglanspollenites nigripites* (Wodehouse) Wingate & Nichols 2001
>
> *Chenopodipollis* sp.

Angiosperm Pollen: Pantoporate, Reticulate

> *Erdtmanipollis procumbentiformis* (Samoilovitch in Samoilovitch & Mtchedlishvili) Krutzsch 1966 [of Wingate and Nichols 2001]
>
> *Persicarioipollis* sp. cf. *P. welzowense* Krutzsch 1962 [of Wingate and Nichols 2001]
>
> ?*Persicarioipollis* sp. A
>
> ?*Persicarioipollis* sp. B

Angiosperm Pollen: Pantoporate, Echinate

> *Malvacipollis* sp. A
>
> *Malvacipollis* sp. B
>
> *Malvacipollis* sp. C
>
> *Malvacipollis* sp. D
>
> ?*Periporopollenites* sp. cf. *P. stigmosus* (Potonié) Pflug & Thomson in Thomson & Pflug 1953 [of Wingate and Nichols 2001]

Angiosperm Pollen: Monoporate

> *Cyperaceaepollis* sp. cf. *C. neogenicus* Krutzsch 1970 [of Wingate and Nichols 2001]
>
> *Graminidites crassiglobosus* (Trevisan) Krutzsch 1970 [of Wingate and Nichols 2001]
>
> ?*Pandaniidites* sp.
>
> *Sparganiaceaepollenites sparganioides* (Meyer)

Krutzsch 1970 [of Wingate and Nichols 2001]

Angiosperm Pollen: Diporate

Diporate pollen cf. *Trema* Lour.

Angiosperm Pollen: Tetrads

Ericipites sp. cf. *E. longisulcatus* Wodehouse 1933 [of Wingate and Nichols 2001]

Angiosperm Pollen: Genus and Species Indeterminate

?Triporate sp. cf. ?*Ruellia laxa* (Frederiksen) Frederiksen et al. 1983 [of Wingate and Nichols 2001]

POLLEN AND SPORES (OF LEOPOLD AND CLAY-POOLE 2001)

Division Lycophyta (lycopods)

ORDER SELAGINELLALES

Family SELAGINELLACEAE

Selaginella

Division Pteridophyta

ORDER FILICALES (ferns)

Family SCHIZAEACEAE

Lygodium

Family POLYPODIACEAE

Polypodicaceae-type spores

Division Gnetophylla

ORDER GNETALES

Family EPHEDRACEAE [GNETACEAE] (joint-fir family)

Ephedra

Division Coniferophyta

ORDER CONIFERALES (conifers)

Families TAXODIACEAE-CUPRESSACEAE-TAXACEAE

TCT pollen type

Family TAXODIACEAE [-CUPRESSACEAE]

Sequoia affinis Lesquereux 1876

Family PINACEAE (pine family)

Abies

Pseudotsuga /*Larix*

Picea spp.

Tsuga spp.

Pinus spp.

Cedrus (extinct type)

Division Magnoliophyta

Class Magnoliopsida (dicotyledonous flowering plants)

ORDER NYMPHAEALES

Family NYMPHAEACEAE (water lily family)

Nuphar

ORDER BERBERIDALES

Family BERBERIDACEAE (barberry family)

Mahonia/*Berberis*

ORDER CARYOPHYLLALES

Family CARYOPHYLLACEAE (pink family)

Stellaria or *Silene*-type

Family CHENOPODIACEAE/AMARANTHACEAE (goosefoot/amaranth families)

Amaranthus or Chenopodiaceae

Sacrobatus

ORDER FAGALES

Family FAGACEAE (beech family)

Castanea

Fagopsis

Fagus

Quercus

Quercoid, long-axial pollen

ORDER CORYLALES

Family BETULACEAE (birch family)

Ostrya-Carpinus (*Ostrya*-type)

ORDER RHOIPTELEALES

Family RHOIPTELEACEAE

aff. *Rhoiptelea*

ORDER JUGLANDALES

Family JUGLANDACEAE (walnut family)

    Carya

    Cyclocarya

    Engelhardtia/Alfaroa

    Juglans

    Pterocarya

    Platycarya

ORDER ERICALES

Family ERICACEAE or PYROLACEAE

    Pollen of Ericaceae or Pyrolaceae

ORDER SALICALES

Family SALICACEAE (willow family)

    Populus

    Salix

ORDER MALVALES

Family STERCULIACEAE (cocoa family)

    Florissantia-type

    Fremontodendreae

Family MALVACEAE (mallow family)

    Malvaceous pollen

ORDER URTICALES

Family ULMACEAE (elm family)

    Pteroceltis

    Zelkova

Family MORACEAE (mulberry family)

    Morus-type

ORDER EUPHORBIALES

Family EUPHORBIACEAE (spurge family)

    Croton

ORDER THYMELAEALES

Family THYMELAEACEAE

    Daphne

ORDER ROSALES

Family ROSACEAE (rose family)

    Pyrus

ORDER MYRTALES

Family ONAGRACEAE (evening primrose family)

    Semeiandra

    aff. *Xylonagra*

ORDER FABALES

Family FABACEAE [= LEGUMINOSAE] (legume family)

    cf. *Petalostemon*

ORDER SAPINDALES

Family SAPINDACEAE (soapberry family)

    Cardiospermum

    Koelreuteria

Family ACERACEAE [~SAPINDACEAE] (maple family)

    Acer

ORDER RUTALES

Family RUTACEAE (citrus family)

    ?*Ptelea*

    aff. *Evodia*

ORDER BURSERALES

Family ANACARDIACEAE (cashew family)

    Astronium-type

ORDER ELAEAGNALES

Family ELAEAGNACEAE (oleaster family)

    Elaeagnus

ORDER VITALES

Family VITACEAE (grape family)

    Parthenocissus/Cissus

ORDER EUCOMMIALES

Family EUCOMMIACEAE (hard-rubber tree family)

    Eucommia

ORDER ADOXALES

Family SAMBUCACEAE (elder family)

    Sambucus

ORDER DIPSACALES

Family CAPRIFOLIACEAE (honeysuckle family)

    Viburnum

ORDER APOCYNALES

Family APOCYNACEAE (sweetsop family)

   Tabernaemontana

ORDER SOLANALES

Family SOLANACEAE (potato family)

   Datura

ORDER OLEALES

Family OLEACEAE (olive family)

   Fraxinus-type

Class Liliopsida (monocotyledonous flowering plants)

ORDER POALES

Family POACEAE [= GRAMINEAE] (grass family)

   Poaceae pollen

ORDER ARECALES

Family ARECACEAE [= PALMAE] (palm family)

   Palmae-type pollen

ORDER TYPHALES

Family TYPHACEAE/SPARGANIACEAE (cattail/bur reed families)

   Typha/Sparganium

Incertae sedis (uncertain)

   Asteraceae/Compositae?

   Aquifoliaceae?

      Ilex-type

   Haloragidaceae?

      Myriophyllum-type

INVERTEBRATES

SPIDERS AND MILLIPEDES

Phylum ARTHROPODA

Class Arachnida (spiders)

ORDER ARANEAE (true spiders)

Family THERAPHOSIDAE (tarantulas)

   †*Eodiplurina cockerelli* Petrunkevitch 1922: NHM, UCM

Family SEGESTRIIDAE (segestriid six-eyed spiders)

   Segestria scudderi Petrunkevitch 1922: MCZ

   Segestria secessa Scudder 1890: MCZ, USNM

Family GNAPHOSIDAE (hunting spiders)

   †*Palaeodrassus cockerelli* Petrunkevitch 1922: UCM

   †*Palaeodrassus florissanti* Petrunkevitch 1922: UCM

   †*Palaeodrassus hesternus* (Scudder) Petrunkevitch 1922: MCZ

   †*Palaeodrassus ingenuus* (Scudder) Petrunkevitch 1922: MCZ

   †*Palaeodrassus interitus* (Scudder) Petrunkevitch 1922: MCZ

Family LINYPHIIDAE (sheet web spiders; bowl and doily spiders)

   Linyphia florissanti Petrunkevitch 1922: UCM

   Linyphia pachygnathoides Petrunkevitch 1922: [specimen has not been located]

   Linyphia retensa Scudder 1890: MCZ

   Linyphia seclusa (Scudder) Petrunkevitch 1922: MCZ

Family ARANEIDAE (orb weavers)

  Subfamily Araneinae

   Araneus abscondita [= *Epeira abscondita* Scudder 1890]: MCZ

   Araneus cinefacta [= *Epeira cinefacta* Scudder 1890]: MCZ

   Araneus delita [= *Epeira delita* Scudder 1890]: MCZ

   Araneus emertoni [= *Epeira emertoni* Scudder 1890]: MCZ

   Araneus indistincta [= *Epeira indistincta* Petrunkevitch 1922]: MCZ

   Araneus longimana [= *Epeira longimana* Petrunkevitch 1922]: MCZ

   Araneus meekii [= *Epeira meeki* Scudder 1890]: MCZ

   Araneus vulcanalis [= *Epeira vulcanalis* Scudder 1890]: MCZ

   †*Tethneus guyoti* Scudder 1890: MCZ, USNM

   †*Tethneus hentzii* Scudder 1890: MCZ

   †*Tethneus obduratus* Scudder 1890: MCZ

†*Tethneus provectus* Scudder 1890: MCZ

†*Tethneus robustus* Petrunkevitch 1922: MCZ

†*Tethneus twenhofeli* Petrunkevitch 1922: YPM

Subfamily Nephilinae (silk spiders)

Nephila pennatipes Scudder 1885: MCZ

Family TETRAGNATHIDAE (long-jawed orb weavers)

Subfamily Tetragnathinae

†*Palaeometa opertanea* (Scudder) Petrunkevitch 1922: MCZ

†*Palaeopachygnatha cockerelli* Petrunkevitch 1922: UCM

†*Palaeopachygnatha scudderi* Petrunkevitch 1922: MCZ

Tetragnatha tertiaria Scudder 1890: MCZ

Family THOMISIDAE (crab spiders)

Thomisus defossus Scudder 1890: MCZ

Thomisus disjunctus Scudder 1890: MCZ

Thomisus resutus Scudder 1890: MCZ

Family CLUBIONIDAE (sac spiders)

Subfamily Clubioninae

Clubiona arcana Scudder 1890: MCZ, USNM

Clubiona curvispinosa Petrunkevitch 1922: MCZ

Clubiona florissanti Petrunkevitch 1922: UCM

†*Eobumbratrix latebrosa* (Scudder) Petrunkevitch 1922: MCZ

†*Eostentatrix cockerelli* Petrunkevitch 1922: UCM

†*Eostentatrix ostentata* (Scudder) Petrunkevitch 1922: MCZ

†*Eoversatrix eversa* (Scudder) Petrunkevitch 1922: MCZ

†Family PARATTIDAE (extinct clubionid-like spiders)

†*Parattus evocatus* Scudder 1890: MCZ

†*Parattus latitatus* Scudder 1890: MCZ

†*Parattus oculatus* Petrunkevitch 1922: MCZ

†*Parattus resurrectus* Scudder 1890: MCZ

Family LYCOSIDAE (wolf spiders)

Lycosa florissanti Petrunkevitch 1922: AMNH

ORDER OPILIONES (harvestmen)

Family LEIOBUNIDAE

Subfamily Leiobuninae

†*Amauropilio atavus* (Cockerell) Cokendolpher & Cokendolpher 1982: AMNH, NHM

†*Amauropilio lacoei* (Petrunkevitch) Mello-Leitao 1937

Family incertae sedis (uncertain)

†*Petrunkevitchiana oculata* (Petrunkevitch) Mello-Leitao 1937 [= *Phalangium* Petrunkevitch 1922]: MCZ

ORDER ACARI (mites and ticks)

Family ERIOPHYIDAE

Eriophyes beutenmülleri Cockerell 1908: NHM, UCM

Class Diplopoda (millipedes)

ORDER JULIDA

Family JULIDAE

Julus florissantellus Cockerell 1907: AMNH

Family PARAJULIDAE

Parajulus cockerelli Miner 1926: AMNH

INSECTS

Phylum ARTHROPODA

Class Hexapoda (insects)

ORDER EPHEMEROPTERA (mayflies)

Family SIPHLONURIDAE

†*Siphlurites explanatus* Cockerell 1923: [specimen has not been located]

Family LEPTOPHLEBIIDAE

†*Lepismophlebia platymera* (Scudder) Demoulin 1968: MCZ

Family EPHEMERIDAE (common and burrowing mayflies)

Subfamily Ephemerinae

Ephemera exsucca Scudder 1890: [specimen has not been located]

Ephemera howarthi Cockerell 1908: AMNH, YPM

Ephemera immobilis Scudder 1890: [specimen has not been located]

Ephemera interment Scudder 1890: [specimen has not been located]

Ephemera macilenta Scudder 1890: [specimen has not been located]

Ephemera pumicosa Scudder 1890: MCZ

Ephemera tabifica Scudder 1890: [specimen has not been located]

ORDER ODONATA (dragonflies and damselflies)

Family AESHNIDAE (darners)

Aeschna larvata Scudder 1890: MCZ

Aeschna solida Scudder 1890: MCZ

Hoplonoaeschna separata (Scudder) Needham 1903 [reference in Cockerell 1908]: MCZ

†*Lithaeschna needhami* Cockerell 1907: UCM

Oplonaeschna lapidaris Cockerell 1913: UCM, USNM

Family CALOPTERYGIDAE (broad-winged damselflies)

Calopteryx telluris [= *Agrion telluris* Scudder 1890]: MCZ

Family COENAGRIONIDAE (narrow-winged damselflies)

Enallagma exsulare (Scudder) Cockerell 1925: MCZ

Enallagma florissantellum Cockerell 1908: YPM

Enallagma mascescens (Scudder) Cockerell 1925: MCZ

Enallagma mortuellum Cockerell 1909: UCM

Enallagma oblisum Cockerell 1925: UCM

Enallagma praevolans (Cockerell) Cockerell 1925: AMNH, UCM

Trichocnemis aliena Scudder 1892: [specimen has not been located]

Family PSEUDOLESTIDAE

†*Phenacolestes mirandus* Cockerell 1908: UCM

†*Phenacolestes*(?) *parallelus* Cockerell 1908: UCM

†Family MEGAPODAGRIONIDAE

†*Lithagrion hyalinum* Scudder 1890: MCZ

†*Melanagrion nigerrimum* Cockerell 1908: YPM

†*Melanagrion umbratum* (Scudder) Cockerell 1907: MCZ

†*Miopodagrion optimum* (Cockerell) Kennedy 1925: UCM

ORDER PHASMIDA (walkingsticks and leaf insects)

Family PHASMATIDAE (winged walkingsticks)

Subfamily Anisomorphinae

Agathemera reclusa Scudder 1890: [specimen has not been located]

ORDER ORTHOPTERA (grasshoppers, crickets, and katydids)

Family EUMASTICIDAE

†*Taphacris bittaciformis* (Cockerell) Cockerell 1926 : UCM

†*Taphacris reliquata* Scudder 1890: MCZ

Family ACRIDIDAE (short-horned grasshoppers)

Subfamily Acridinae (slant-faced grasshoppers)

Gomphocerus abstrusus Scudder 1890: MCZ

†*Taeniopodites pardalis* Cockerell 1909: NHM, UCM

†*Tyrbula multispinosa* Scudder 1890: MCZ

†*Tyrbula russelli* Scudder 1885: MCZ

†*Tyrbula scudderi* Cockerell 1914: USNM

Subfamily Oedipodinae

†*Nanthacia torpida* Scudder 1890: MCZ

Oedipoda praefocata Scudder 1890: MCZ

Family TETTIGONIIDAE (long-horned grasshoppers)

Subfamily Phyllophorinae

†*Lithymnetes guttatus* Scudder 1878: MCZ

Subfamily Phaneropterinae (katydids)

Amblycorypha perdita Cockerell 1914: USNM

Subfamily Conocephalinae (meadow grasshoppers)

Locusta silens Scudder 1890: MCZ

Orchelimum placidum Scudder 1890: MCZ

Subfamily Tettigoniinae (shield-back grasshoppers and pine tree katydids)

Anabrus caudelli Cockerell 1908: UCM

Family GRYLLACRIDIDAE (wingless long-horned grasshoppers)

   Gryllacris cineris Scudder 1890: MCZ

   Gryllacris mutilata Cockerell 1909: UCM

Family HAGLIDAE

   †*Palaeorehnia maculata* (Scudder) Cockerell 1908: NHM, UCM

Family GRYLLIDAE (crickets)

   Subfamily Gryllinae

      †*Lithogryllites lutzii* Cockerell 1908: AMNH, UCM

ORDER MANTODEA (mantids)

Family MANTIDAE (mantids)

   †*Eobruneria tessellata* Cockerell 1913: USNM

Family CHAETEESSIDAE

   †*Lithophotina costalis* Cockerell 1914: USNM

   †*Lithophotina floccosa* Cockerell 1908: NHM, UCM

ORDER BLATTARIA (cockroaches)

Family BLATTELIDAE (German and wood cockroaches)

   Ischnoptera brunneri Cockerell 1909: [specimen has not been located]

   Zetobora brunneri Scudder 1890: MCZ

Family HOMOEOGAMIIDAE

   Homoeogamia ventriosus Scudder 1876: MCZ

ORDER ISOPTERA (termites)

Family KALOTERMITIDAE (drywood, dampwood, and powderpost termites)

   †*Proelectrotermes fodinae* (Scudder) Emerson 1969: MCZ

   †*Prokalotermes hagenii* (Scudder) Emerson 1933: MCZ

Family HODOTERMITIDAE (dampwood termites)

   Subfamily Termopsinae

      †*Parotermes insignis* Scudder 1884: MCZ

   Zootermopsis(?) *coloradensis* (Scudder) Emerson 1933: MCZ

Family RHINOTERMITIDAE (subterranean and dampwood termites)

   Reticulitermes fossarum (Scudder) Snyder 1925: MCZ

ORDER DERMAPTERA (earwigs)

Family LABIDURIDAE (striped earwigs)

   †*Labiduromma avia* Scudder 1885: MCZ

   †*Labiduromma bormansi* Scudder 1890: MCZ, USNM

   †*Labiduromma commixtum* Scudder 1890: MCZ

   †*Labiduromma exsulatum* Scudder 1885: MCZ

   †*Labiduromma gilberti* Scudder 1890: MCZ

   †*Labiduromma gurneyi* Brown 1984: UCM

   †*Labiduromma infernum* Scudder 1890: MCZ

   †*Labiduromma labens* Scudder 1890: MCZ

   †*Labiduromma lithophilum* (Scudder) Scudder 1890: MCZ

   †*Labiduromma mortale* Scudder 1890: MCZ, USNM

   †*Labiduromma scudderi* Brown 1984: UCM

   †*Labiduromma tertiarium* (Scudder) Scudder 1890: USNM

ORDER EMBIIDINA (web-spinners)

Family EMBIIDAE

   †*Lithembia florissantensis* (Cockerell) Ross 1984: UCM, YPM

ORDER HEMIPTERA (true bugs)

Family NEPIDAE (waterscorpions)

   Nepa vulcanica (Cockerell) Hungerford 1932: NHM, YPM

Family CORIXIDAE (water boatmen)

   Corixa immersa Scudder 1890: MCZ

   Corixa vanduzeei Scudder 1890: MCZ

   †*Sigaretta florissantella* (Cockerell) Popov 1971: NHM

Family NOTONECTIDAE (backswimmers)

   Notonecta emersoni Scudder 1890: MCZ

Family VELIIDAE (broad-shouldered water striders, riffle bugs)

   †*Palaeovelia spinosa* Scudder 1890: MCZ

   †*Stenovelia nigra* Scudder 1890: MCZ

Family GERRIDAE (water striders)
    Gerris protobates Cockerell 1927: NHM
    Metrobates aeternalis Scudder 1890: MCZ

Family TINGIDAE (lace bugs)
    Dictyla veterna (Scudder) Drake & Ruhoff 1960: MCZ
    †*Eotingis antennata* Scudder 1890: MCZ
    Tingis florissantensis Cockerell 1914: UCM
    Tingis sp. Scudder 1881: [specimen has not been located]

Family MIRIDAE (leaf bugs, plant bugs)
    †*Aporema praestrictum* Scudder 1890: MCZ
    Capsus(?) *lacus* Scudder 1890: MCZ
    Capsus(?) *obsolefactus* Scudder 1890: [specimen has not been located]
    Carmelus(?) *gravatus* Scudder 1890: MCZ
    Carmelus(?) *sepositus* Scudder 1890: MCZ
    Closterocoris(?) *elegans* Scudder 1890: MCZ
    Fuscus(?) *faecatus* Scudder 1890: MCZ
    Hadronema cinerescens Scudder 1890: MCZ
    Poecilocapsus fremontii Scudder 1890: MCZ, USNM
    Poecilocapsus ostentus Scudder 1890: MCZ
    Poecilocapsus tabidus Scudder 1890: MCZ
    Poecilocapsus veterandus Scudder 1890: MCZ
    Poecilocapsus veternosus Scudder 1890: MCZ

Family REDUVIIDAE (assassin bugs, ambush bugs, thread-legged bugs)
    †*Eothes elegans* Scudder 1890: MCZ
    †*Miocoris fagi* Cockerell 1927: NHM
    †*Poliosphageus psychrus* Kirkaldy 1910: UCM
    †*Tagalodes inermis* Scudder 1890: MCZ

Family PIESMATIDAE (ash-gray leaf bugs)
    Subfamily Piesmatinae
    Piesma rotunda Scudder 1890: MCZ

Family LYGAEIDAE (seed bugs)
    †*Catopamera augheyi* Scudder 1890: MCZ
    †*Catopamera bradleyi* Scudder 1890: MCZ
    †*Cophocoris tenebricosus* Scudder 1890: MCZ
    †*Coptochromus manium* Scudder 1890: MCZ
    †*Cryptochromus letatus* Scudder 1890: MCZ
    †*Ctereacoris primigenius* Scudder 1890: MCZ
    †*Eucorites serescens* Scudder 1890: MCZ
    †*Exitelus exsanguis* Scudder 1890: MCZ
    Geocoris infernorum Scudder 1890: MCZ
    Ligyrocoris exsuctus Scudder 1890: MCZ
    †*Linnaea abolita* Scudder 1890: MCZ
    †*Linnaea carcerata* Scudder 1890: MCZ
    †*Linnaea evoluta* Scudder 1890: MCZ
    †*Linnaea gravida* Scudder 1890: MCZ
    †*Linnaea holmseii* Scudder 1890: MCZ
    †*Linnaea putnami* Scudder 1890: MCZ
    †*Lithochromus extraneus* Scudder 1890: MCZ
    †*Lithochromus gardnerii* Scudder 1890: MCZ
    †*Lithochromus mortuarius* Scudder 1890: MCZ
    †*Lithochromus obstrictus* Scudder 1890: MCZ
    †*Lithocoris evulsus* Scudder 1890: MCZ
    Lygaeus faeculentus Scudder 1890: USNM
    Lygaeus obsolescens Scudder 1890: MCZ
    Lygaeus stabilitus Scudder 1890: MCZ, USNM
    †*Necrochromus cockerelli* Scudder 1890: MCZ
    †*Necrochromus labatus* Scudder 1890: MCZ
    †*Necrochromus saxificus* Scudder 1890: MCZ
    Nysius stratus Scudder 1890: USNM
    Nysius terrae Scudder 1890: MCZ
    Nysius tritus Scudder 1890: MCZ
    Nysius vecula Scudder 1890: USNM
    Nysius vinctus Scudder 1890: MCZ
    †*Phrudopamera chittendeni* Scudder 1890: MCZ
    †*Phrudopamera wilsoni* Scudder 1890: MCZ
    †*Procoris bechleri* Scudder 1890: MCZ
    †*Procoris sanctaejohannis* Scudder 1890: MCZ
    †*Procrophius communis* Scudder 1890: MCZ
    †*Procrophius costalis* Scudder 1890: MCZ
    †*Procrophius languens* Scudder 1890: MCZ
    Procymus cockerelli Usinger 1940: CAS, MCZ
    †*Prolygaeus inundatus* Scudder 1890: MCZ

Rhyparochromus verrillii Scudder 1890: MCZ

†*Stenopamera subterrea* Scudder 1890: MCZ

†*Stenopamera tenebrosa* Scudder 1890: MCZ

†*Tiromerus tabifluus* Scudder 1890: MCZ

†*Tiromerus torpefactus* Scudder 1890: MCZ

Trapezonotus exterminatus Scudder 1890: MCZ

Trapezonotus stygialis Scudder 1890: MCZ

Family PYRRHOCORIDAE (red bugs and cotton stainers)

Dysdercus cinctus Scudder 1890: MCZ

Dysdercus unicolor Scudder 1890: MCZ

Family COREIDAE (squash bugs, leaf-footed bugs)

†*Achrestocoris cinerarius* Scudder 1890: MCZ

Anasa priscoputida Scudder 1890: MCZ

Corizus abditivus Scudder 1890: MCZ

Corizus celatus Scudder 1890: MCZ

Corizus somnurnus Scudder 1890: MCZ

Heeria foeda Scudder 1890: MCZ

Heeria gulosa Scudder 1890: MCZ, USNM

Heeria lapidosa Scudder 1890: MCZ, USNM

†*Phthinocoris colligatus* Scudder 1890: MCZ

†*Phthinocoris languidus* Scudder 1890: MCZ

†*Phthinocoris lethargicus* Scudder 1890: MCZ

†*Phthinocoris petraeus* Scudder 1890: MCZ

†*Piezocoris compactilis* Scudder 1890: MCZ

†*Piezocoris peremptus* Scudder 1890: MCZ

†*Piezocoris peritus* Scudder 1890: MCZ

Family ALYDIDAE (broad-headed bugs)

Subfamily Leptocorisinae

†*Orthriocorisa longipes* Scudder 1890: MCZ

Subfamily Micrelytrinae

†*Cydamus*(?) *robustus* Scudder 1890: MCZ

Protenor(?) *imbecilis* Scudder 1890: MCZ

Family CYDNIDAE (burrower bugs)

†*Discostoma* sp. Scudder 1890: [specimen has not been located]

†*Necrocydnus amyzonus* Scudder 1890: MCZ

†*Necrocydnus revectus* Scudder 1890: MCZ

†*Necrocydnus senior* Scudder 1890: MCZ

†*Necrocydnus solidatus* Scudder 1890: MCZ

†*Necrocydnus stygius* Scudder 1890: MCZ

†*Necrocydnus torpens* Scudder 1890: MCZ

†*Necrocydnus vulcanius* Scudder 1890: MCZ

†*Procydnus devictus* Scudder 1890: MCZ

†*Procydnus divexus* Scudder 1890: MCZ

†*Procydnus eatoni* Scudder 1890: MCZ

†*Procydnus mamillanus* Scudder 1890: [type not from Florissant]

†*Procydnus pronus* Scudder 1890: MCZ

†*Procydnus quietus* Scudder 1890: MCZ

†*Procydnus reliquus* Scudder 1890: MCZ

†*Procydnus vesperus* Scudder 1890: MCZ

†*Thlibomenus limosus* Scudder 1890: MCZ

†*Thlibomenus macer* Scudder 1890: MCZ

†*Thlibomenus parvus* Scudder 1890: MCZ

†*Thlibomenus perennatus* Scudder 1890: MCZ

†*Thlibomenus petraeus* Scudder 1890: MCZ

Family PENTATOMIDAE (stink bugs)

†*Cacoschistus maceriatus* Scudder 1890: MCZ

Mecocephala sp. Scudder 1890: [specimen has not been located]

†*Pentatomites folarium* Scudder 1890: MCZ

†*Poliocoris amnesis* Kirkaldy 1910: UCM

†*Polioschistus lapidarius* Scudder 1890: MCZ

†*Polioschistus ligatus* Scudder 1890: MCZ

†*Poteschistus obnubilus* Scudder 1890: MCZ

†*Teleocoris pothetias* Kirkaldy 1910: UCM

†*Teleoschistus placatus* Scudder 1890: MCZ

†*Teleoschistus rigoratus* Scudder 1890: MCZ

†*Thlimmoschistus gravidatus* Scudder 1890: MCZ

†*Thnetoschistus revulsus* Scudder 1890: USNM

†*Tiroschistus indurescens* Scudder 1890: MCZ

Family incertae sedis

†*Cacalydus exstirpatus* Scudder 1890: MCZ

†*Cacalydus lapsus* Scudder 1890: USNM

†*Docimus psylloides* Scudder 1890: MCZ

†*Etirocoris infernalis* Scudder 1890: MCZ

†*Prosigara flabellum* Scudder 1890: MCZ

†*Rhepocoris abscissus* (Scudder) Štys & Říha 1977: MCZ

†*Rhepocoris caducus* (Scudder) Štys & Říha 1977: MCZ

†*Rhepocoris collisus* (Scudder) Štys & Říha 1977: MCZ

†*Rhepocoris defectus* (Scudder) Štys & Říha 1977: MCZ

†*Rhepocoris exanimatus* (Scudder) Štys & Říha 1977: MCZ

†*Rhepocoris inhibitus* (Scudder) Štys & Říha 1977: MCZ

†*Rhepocoris macrescens* (Scudder) Štys & Říha 1977: MCZ

†*Rhepocoris minimus* (Scudder) Štys & Říha 1977: MCZ

†*Rhepocoris praetectus* (Scudder) Štys & Říha 1977: MCZ

†*Rhepocoris praevalens* Scudder 1890: MCZ, USNM

†*Rhepocoris propinquans* Scudder 1890: MCZ

†*Tenor speluncae* Scudder 1890: MCZ

ORDER HOMOPTERA (cicadas, hoppers, psyllids, whiteflies, aphids, and scale insects)

Family CICADIDAE (cicadas)

†*Lithocicada perita* Cockerell 1906: AMNH, NHM

Platypedia primigenia Cockerell 1908: YPM

Tibicen grandiosa (Scudder) Cooper 1941: [specimen has not been located]

Family CERCOPIDAE (froghoppers, spittlebugs)

Aphrophora sp. Scudder 1890: MCZ

Cercopis suffocata Scudder 1890: MCZ

Clastoptera comstocki Scudder 1890: MCZ

†*Palaphrodes cincta* Scudder 1890: MCZ, USNM

†*Palaphrodes irregularis* Scudder 1890: MCZ, USNM

†*Palaphrodes obliqua* Scudder 1890: MCZ

†*Palaphrodes obscura* Scudder 1890: MCZ, USNM

†*Palaphrodes transversa* Scudder 1890: MCZ

†*Palecphora communis* Scudder 1890: MCZ, USNM

†*Palecphora inornata* Scudder 1890: MCZ

†*Palecphora maculata* Scudder 1890: MCZ

†*Palecphora marvinei* Scudder 1890: MCZ

†*Palecphora praevalens* Scudder 1890: MCZ, USNM

†*Petrolystra gigantea* Scudder 1878: MCZ

†*Petrolystra heros* Scudder 1878: MCZ

Prinecphora balteata Scudder 1890: MCZ

Family CICADELLIDAE (leafhoppers)

Acocephalus callosus Scudder 1890: MCZ

Agallia abstructa Scudder 1890: MCZ

Agallia flaccida Scudder 1890: MCZ

Agallia instabilis Scudder 1890: MCZ

Agallia lewisii Scudder 1890: MCZ, USNM

Gypona(?) *cinercia* Scudder 1890: MCZ

†*Jassopsis evidens* Scudder 1890: MCZ

Jassus latebrae Scudder 1890: MCZ

Tettigella priscotincta (Scudder) Statz 1950: MCZ

Thamnotettix fundi Scudder 1890: MCZ

Family CIXIIDAE (cixiid planthoppers)

Cixius(?) *proavus* Scudder: USNM

†*Diaplegma abductum* Scudder 1890: MCZ

†*Diaplegma haldemani* Scudder 1890: MCZ

†*Diaplegma occultorum* Scudder 1890: MCZ

†*Diaplegma ruinosum* Scudder 1890: MCZ

†*Diaplegma venerabile* Scudder 1890: MCZ

†*Diaplegma veterascens* Scudder 1890: MCZ

Family DICTYOPHARIDAE (dictyophard planthoppers)

Dictyophara bouvei Scudder 1890: MCZ

†*Florissantia elegans* Scudder 1890: USNM

Family FULGORIDAE (fulgorid planthoppers)

Fulgora obticescens Scudder 1890: MCZ

†*Nyktalos uhleri* (Scudder) Metcalf 1952: MCZ

†*Nyktalos vigil* (Scudder) Metcalf 1952: MCZ

Family ACHILIDAE (achilid planthoppers)

Elidiptera regularis Scudder 1890: MCZ

Family PSYLLIDAE (jumping plantlice)
 †*Catopsylla prima* Scudder 1890: MCZ
 †*Necropsylla rigida* Scudder 1890: MCZ
 †*Necropsylla rigidula* Cockerell 1911: UCM
 †*Psyllites crawfordi* Cockerell 1914: MCZ, UCM

Family DREPANOSIPHIDAE
 †*Aphidopsis margarum* Scudder 1890: MCZ
 †*Siphonophoroides antiqua* Buckton 1883: MCZ
 †*Siphonophoroides lassa* (Scudder) Heie 1967: MCZ
 †*Siphonophoroides pennata* (Buckton) Heie 1967: MCZ
 †*Siphonophoroides simplex* Buckton 1883: MCZ

Family MINDARIDAE
 Mindarus recurvus (Buckton) Heie 1967: MCZ
 Mindarus scudderi (Buckton) Heie 1967: MCZ

Family COCCIDAE OR DACTLOPIIDAE (soft, wax, and tortoise scale insects)
 Monophlebus simplex Scudder 1890: MCZ

Family incertae sedis
 †*Anconatus dorsuosus* Buckton 1883: MCZ
 †*Anconatus gillettei* Cockerell 1908: NHM, UCM
 †*Anconatus niger* (Scudder) Heie 1967: MCZ
 †*Echinaphis rohweri* Cockerell 1913: UCM
 †*Lithecphora diaphana* Scudder 1890: MCZ
 †*Lithecphora murata* Scudder 1890: MCZ
 †*Lithecphora setigera* Scudder 1890: USNM
 †*Lithecphora unicolor* Scudder 1890: MCZ
 †*Locrites copei* Scudder 1890: MCZ, USNM
 †*Locrites whitei* Scudder 1890: MCZ

ORDER NEUROPTERA (alderflies, dobsonflies, fishflies, snakeflies, lacewings, antlions, and owlflies)

Family RAPHIDIIDAE (raphidiid snakeflies)
 Raphidia elegans (Cockerell) Carpenter 1936: AMNH, UCM
 Raphidia exhumata Cockerell 1909: UCM
 Raphidia mortua Rohwer 1909: UCM
 Raphidia tranquilla Scudder 1890: MCZ
 Raphidia tumulata (Scudder) Carpenter 1936: MCZ
 Raphidia veterana (Scudder) Carpenter 1936: USNM

Family OSMYLIDAE (snakeflies)
 †*Lithosmylus columbianus* (Cockerell) Carpenter 1943: UCM
 †*Osmylidia requieta* (Scudder) Cockerell 1908: MCZ

Family INOCELLIIDAE (inocelliid snakeflies)
 Fibla exusta (Cockerell & Custer) Carpenter 1936: UCM

Family CHRYSOPIDAE (common lacewings, green lacewings)
 †*Archaeochrysa fracta* (Cockerell) Adams 1967: MCZ
 †*Archaeochrysa paranervis* Adams 1967: UCM
 †*Dyspetochrysa vetuscula* (Scudder) Adams 1967: MCZ
 †*Palaeochrysa concinnula* Cockerell 1909: UCM
 †*Palaeochrysa stricta* Scudder 1890: MCZ
 †*Palaeochrysa wickhami* Cockerell 1914: MCZ
 †*Tribochrysa firmata* Scudder 1890: MCZ
 †*Tribochrysa inaequalis* Scudder 1885: MCZ

Family POLYSTOECHOTIDAE (giant lacewings)
 Polystoechotes piperatus Cockerell 1908: AMNH

Family NEMOPTERIDAE
 †*Marquettia americana* (Cockerell) Navas 1913: AMNH, NHM

ORDER COLEOPTERA (beetles)

Family CARABIDAE (ground beetles)
 Subfamily Paussinae
  †*Paussopsis*(?) *nearctica* Cockerell 1911: UCM
  †*Paussopsis*(?) *secunda* Wickham 1912: UCM
 Subfamily Carabinae
  "*Calosoma*" *calvini* Wickham 1909: YPM
  "*Calosoma*" *cockerelli* Wickham 1910: YPM
  "*Calosoma*" *emmonsii* Scudder 1900: MCZ
  "*Carabus*" *jeffersoni* Scudder 1900: MCZ
  "*Nebria*" *occlusa* Scudder 1900: MCZ

"Scaphinotus" serus [= "Nomaretus" serus Scudder 1900]: MCZ

Subfamily Trechinae

"Bembidion" florissantensis Wickham 1913: USNM

"Bembidion" obductum Scudder 1900: MCZ

"Bembidion" tumulorum Scudder 1900: USNM

"Tachys" haywardi Wickham 1913: USNM

"Trechus" fractus Wickham 1913: UCM

Subfamily Harpalinae

"Amara" cockerelli Wickham 1912: UCM

"Amara" danae Scudder 1900: MCZ, USNM

"Amara" powellii Scudder 1900: MCZ

"Amara" revocata Scudder 1900: MCZ

"Amara" sterilis Scudder 1900: MCZ, USNM

"Amara" veterata Scudder 1900: MCZ

"Cratacanthus" florissantensis Wickham 1917: USNM

"Cyclotrachelus" tenebricus [= "Evarthrus" tenebricus Scudder 1900]: MCZ

Diplocheila henshawi Scudder 1890: MCZ

"Euryderus" kingii [= "Nothopus" kingii Scudder 1900]: MCZ

"Harpalus" maceratus Wickham 1911: UCM

"Harpalus" nuperus Scudder 1900: MCZ

"Harpalus" redivivus Wickham 1917: USNM

"Harpalus" ulomaeformis Wickham 1917: USNM

"Harpalus" whitfieldi Scudder 1900: MCZ, USNM

"Myas" rigefactus Scudder 1900: MCZ

"Myas" umbrarum Scudder 1900: MCZ

"Platynus" florissantensis Wickham 1913: USNM

"Platynus" insculptipennis Wickham 1917: USNM

"Platynus" tartareus Scudder 1900: MCZ

"Plochionus" lesquereuxi Scudder 1900: MCZ

"Pterostichus" pumpellyi Scudder 1900: MCZ

"Pterostichus" walcotti Scudder 1900: MCZ, USNM

"Stenolophus" religatus Scudder 1900: MCZ

Subfamily Brachininae

"Brachinus" newberryi Scudder 1900: MCZ

"Brachinus" repressus Scudder 1900: MCZ

Family DYTISCIDAE (predaceous diving beetles)

Subfamily Hydroporinae

"Bidessus" laminarum Wickham 1914: MCZ

"Hydroporus" sedimentorum Wickham 1914: MCZ

"Hygrotus" miocenus [= "Coelambus" miocenus Wickham 1912]: AMNH

Subfamily Colymbetinae

"Agabus" charon Wickham 1912: UCM

"Agabus" florissantensis Wickham 1913: USNM

"Agabus" rathbuni Scudder 1900: MCZ

Subfamily Dytiscinae

"Acilius" florissantensis Wickham 1909: YPM

†Miodytiscus hirtipes Wickham 1911: AMNH

Family AGYRTIDAE (agyritid beetles)

"Agyrtes" primoticus Scudder 1900: MCZ

Family LEIODIDAE (round fungus beetles)

Subfamily Leiodinae

Hydnobius tibialis Wickham 1913: USNM

Family SILPHIDAE (carrion beetles)

Anistoma sibylla Wickham 1913: USNM

Subfamily Silphinae (carrion beetles)

†Miosilpha necrophiloides Wickham 1912: UCM

"Necrodes" primaevus Beutenmüller & Cockerell 1908: AMNH

"Silpha" beutenmuelleri Wickham 1914: MCZ

"Silpha" colorata Scudder 1900: MCZ

Family STAPHYLINIDAE (rove beetles)

†Aleocharopsis caseyi Wickham 1913: USNM

†Aleocharopsis secunda Wickham 1913: USNM

"Geodromicus" abditus Scudder 1900: MCZ

†Laasbium agassizii Scudder 1900: MCZ

†Laasbium sectile Scudder 1900: MCZ

†Miolithocharis lithographica Wickham 1913: USNM

"Omalium" antiquorum [= Homalium antiquorum Wickham 1913]: USNM

†Trigites coeni (Scudder) Handlirsch 1907: MCZ

Subfamily Tachyporinae

"Bolitobius" durabilis Scudder 1900: MCZ

"*Bolitobius*" *funditus* Scudder 1900: MCZ

"*Bolitobius*" *lyelli* Scudder 1900: MCZ

"*Bolitobius*" *stygis* Scudder 1900: MCZ

"*Mycetoporus*" *demersus* Scudder 1900: MCZ

"*Tachinus*" *sommatus* Scudder 1900: MCZ

"*Tachyporus*" *nigripennis* Scudder 1900: MCZ

Subfamily Aleocharinae

"*Atheta*" *florissantensis* Wickham 1913: USNM

Subfamily Oxytelinae

"*Bledius*" *osborni* Scudder 1900: MCZ, USNM

"*Bledius*" *primitiarum* Scudder 1900: MCZ

"*Bledius*" *soli* Scudder 1900: MCZ

"*Deleaster*" *grandiceps* Wickham 1912: UCM

"*Oxytelus*" *subapterus* Wickham 1913: USNM

"*Platystethus*" *archetypus* Scudder 1900: MCZ

"*Platystethus*" *carcareus* Scudder 1900: MCZ

Subfamily Steninae

Stenus morsei (Scudder) Wickham 1913: MCZ, USNM

Subfamily Paederinae

"*Lathrobium*" *antediluvianum* Wickham 1913: USNM

"*Lithocharis*" *scottii* Scudder 1900: USNM

"*Paederus*" *adumbratus* Wickham 1913: USNM

Subfamily Staphylininae

"*Acylophorus*" *immotus* Scudder 1900: MCZ

"*Heterothops*" *contincens* Scudder 1900: MCZ

"*Leptacinus*" *exsucidus* Scudder 1900: MCZ

"*Leptacinus*" *fossus* Scudder 1900: MCZ

"*Leptacinus*" *leidyi* Scudder 1900: MCZ

"*Leptacinus*" *maclurei* Scudder 1900: MCZ

"*Leptacinus*" *rigatus* Scudder 1900: MCZ

"*Philonthus*" *abavus* Scudder 1900: MCZ, USNM

"*Philonthus*" *horni* Scudder 1900: MCZ

"*Philonthus*" *invelatus* Scudder 1900: MCZ

"*Philonthus*" *marcidulus* Scudder 1900: MCZ

"*Quedius*" *breweri* Scudder 1890: MCZ, USNM

"*Quedius*" *chamberlini* Scudder 1890: MCZ

"*Quedius*" *mortuus* Wickham 1912: AMNH

"*Staphylinus*" *lesleyi* Scudder 1900: MCZ, USNM

"*Staphylinus*" *vetulus* Scudder 1900: MCZ

"*Staphylinus*" *vulcan* Wickham 1913: USNM

"*Xantholinus*" *tenebrarius* Scudder 1900: USNM

Family HYDROPHILIDAE (water scavenger beetles)

†*Creniphilites orpheus* Wickham 1913: USNM

"*Philhydrus*" *scudderi* Wickham 1909: YPM

Subfamily Hydrophilinae

"*Hydrobius*" *maceratus* Scudder 1900: MCZ

"*Hydrobius*" *prisconatator* Wickham 1911: AMNH, UCM

"*Hydrobius*" *titan* Wickham 1913: USNM

Hydrochara extricatus Scudder 1900: MCZ

"*Tropisternus*" *limitatus* Scudder 1900: MCZ

"*Tropisternus*" *vanus* Scudder 1900: MCZ

Family SCIRTIDAE (marsh beetles)

†*Miocyphon punctulatus* Wickham 1914: MCZ

Family DASCILLIDAE (soft-bodied plant beetles)

†*Protacnaeus tenuicornis* Wickham 1914: MCZ

†*Scaptolenopsis lithographus* (Wickham) Wickham 1920: USNM

Family LUCANIDAE (stag beetles)

Subfamily Syndesinae

"*Ceruchus*" *fuchsii* Wickham 1911: UCM

Subfamily Lucaninae

"*Lucanus*" *fossilis* Wickham 1913: USNM

Family TROGIDAE (skin beetles)

Trox antiquus Wickham 1909: NHM

Family SCARABAEIDAE (scarab beetles)

"*Amphicoma*" *defuncta* Wickham 1910: YPM

Subfamily Aphodiinae

"*Aphodius*" *aboriginalis* Wickham 1912: UCM

"*Aphodius*" *florissantensis* Wickham 1911: AMNH

"*Aphodius*" *granarioides* Wickham 1913: USNM

"*Aphodius*" *inundatus* Wickham 1914: USNM

"*Aphodius*" *laminicola* Wickham 1910: YPM, UCM

"*Aphodius*" *mediaevus* Wickham 1914: MCZ

"*Aphodius*" *praeemptor* Wickham 1913: USNM

"Aphodius" restructus Wickham 1912: UCM

"Aphodius" senex Wickham 1914: MCZ

"Aphodius" shoshonis Wickham 1912: UCM

"Ataenius" patescens Scudder 1900: MCZ

"Oxyomus" nearcticus Wickham 1914: MCZ

Subfamily Melolonthinae

"Diplotaxis" aurora Wickham 1913: USNM

"Diplotaxis" simplicipes 1912: USNM

"Hoplia" striatipennis Wickham 1914: MCZ

"Macrodactylus" pluto Wickham 1912: UCM

"Macrodactylus" propheticus Wickham 1912: UCM

†*Miuluchnosterna tristoides* Wickham 1914: MCZ

"Phyllophaga" disrupta Cockerell 1927: NHM

"Phyllophaga"(?) extincta Wickham 1916: NHM, UCM

"Phyllophaga" puerilis Wickham 1914: MCZ

"Serica" antediluviana Wickham 1912: UCM

"Serica" cockerelli Wickham 1914: USNM

Subfamily Rutelinae

"Anomala" exterranea Wickham 1914: MCZ

"Anomala" scudderi Wickham 1914: MCZ

Subfamily Dynastinae

"Ligyrus" compositus Wickham 1911: UCM

"Ligyrus" effectus Wickham 1914: MCZ

"Strategus" cessatus Wickham 1914: MCZ

Family BYRRHIDAE (pill beetles)

†*Nosotetocus debilis* Scudder 1900: MCZ

†*Nosotetocus marcovi* Scudder 1892: MCZ

†*Nosotetocus vespertinus* Scudder 1900: MCZ

Subfamily Amphicyrtinae

"Amphicyrta" inhaesa Scudder 1900: MCZ

Subfamily Byrrhinae

"Byrrhus" romingeri Scudder 1900: MCZ, USNM

"Cytilus" dormiscens Scudder 1900: MCZ

"Cytilus" tartarinus Scudder 1900: MCZ

Family BUPRESTIDAE (metallic wood-boring beetles)

†*Brachyspathus curiosus* Wickham 1917: UCM

Subfamily Buprestinae

"Anthaxia" exhumata Wickham 1913: USNM

"Buprestis" florissantensis Wickham 1914: MCZ

"Buprestis" megistarche Cockerell 1926: NHM

"Buprestis" scudderi Wickham 1914: MCZ

"Chrysobothris" coloradensis Wickham 1914: MCZ

"Chrysobothris" gahani Cockerell 1911: UCM

"Chrysobothris" haydeni Scudder 1876: MCZ

"Chrysobothris" suppressa Wickham 1914: MCZ

"Dicerca" eurydice Wickham 1914: MCZ

"Melanophila" cockerellae Wickham 1912: UCM

"Melanophila" handlirschi Wickham 1912: AMNH

"Melanophila" heeri Wickham 1914: USNM

"Ptosima" abyssa (Wickham) Wickham 1914: UCM

"Ptosima" schaefferi (Wickham) Wickham 1914: UCM

"Ptosima" silvatica Wickham 1914: MCZ

Subfamily Argilinae

"Agrilus" praepolitus Wickham 1914: MCZ

Family CHELONARIIDAE (chelonariid beetles)

"Chelonarium" montanum Wickham 1914: MCZ

Family DRYOPIDAE (long-toed water beetles)

"Helichus" eruptus (Wickham) Wickham 1920: AMNH

"Helichus" tenuior (Wickham) Wickham 1920: UCM

†*Lutrochites lecontei* Wickham 1912: AMNH

Family PSEPHENIDAE (water-penny beetles)

Subfamily Eubriinae

"Ectopria" laticollis Wickham 1913: USNM

Subfamily Psepheninae

"Psephenus" lutulenus Scudder 1900: MCZ

Family ELATERIDAE (click beetles)

†*Cryptagriotes minusculus* Wickham 1916: MCZ

"Cryptohypnus" exterminatus Wickham 1916: MCZ

"Cryptohypnus" hesperus Wickham 1916: MCZ

†*Ludiophanes haydeni* Wickham 1916: USNM

"Monocrepidius" dubiosus Wickham 1916: USNM

Subfamily Agrypninae

"Lacon" exhumatus Wickham 1916: MCZ

Subfamily Denticollinae

"*Athous*" *contusus* Wickham 1916: MCZ

"*Athous*" *fractus* Wickham 1916: UCM

"*Athous*" *lethalis* Wickham 1916: MCZ

"*Limonius*" *aboriginalis* Wickham 1916: USNM

"*Limonius*" *florissantensis* Wickham 1916: USNM

"*Limonius*" *praecursor* Wickham 1916: MCZ

"*Limonius*" *shoshonis* Wickham 1916: UCM

"*Limonius*" *volans* Wickham 1916: UCM

"*Melanactes*" *cockerelli* Wickham 1908: YPM

Subfamily Elaterinae

"*Agriotes*" *comminutus* Wickham 1916: MCZ

"*Agriotes*" *nearcticus* Wickham 1916: MCZ

"*Anchastus*" *diluvialis* Wickham 1916: MCZ

"*Anchastus*" *eruptus* Wickham 1916: MCZ

"*Elater*" *exanimatus* [= "*Ludius*" *exanimatus* (Wickham) Wickham 1920]: USNM

"*Elater*" *florissantensis* Wickham 1916: MCZ

"*Elater*" *granulicollis* [= "*Ludius*" *granulicollis* (Wickham) Wickham 1920]: YPM

"*Elater*" *heeri* [= "*Ludius*" *heeri* (Wickham) Wickham 1920]: UCM

"*Elater*" *laevissimus* [= "*Ludius*" *laevissimus* (Wickham) Wickham 1920]: UCM

"*Elater*" *primitivus* [= "*Ludius*" *primitivus* (Wickham) Wickham 1920]: YPM

"*Elater*" *propheticus* [= "*Ludius*" *propheticus* (Wickham) Wickham 1920]: MCZ

"*Elater*" *restructus* [= "*Ludius*" *restructus* (Wickham) Wickham 1920]: UCM

"*Elater*" *rohweri* Wickham 1916: UCM

"*Elater*" *scudderi* Wickham 1916: MCZ

"*Elater*" *submersus* [= "*Ludius*" *submersus* (Wickham) Wickham 1920]: UCM

"*Megapenthes*" *primaevus* Wickham 1916: MCZ

"*Oxygonus*" *primus* Wickham 1916: MCZ

Subfamily Cardiophorinae

"*Cardiophorus*" *cockerelli* Wickham 1916: MCZ

"*Cardiophorus*"(?) *deprivatus* Wickham 1916: UCM

"*Cardiophorus*" *florissantensis* Wickham 1916: UCM

"*Cardiophorus*" *lithographus* Wickham 1916: USNM

"*Cardiophorus*" *requiescens* Wickham 1916: USNM

"*Horistonotus*" *coloradensis* Wickham 1916: USNM

Family THROSCIDAE (throscid beetles)

"*Pactopus*" *americanus* Wickham 1914: USNM

Family EUCNEMIDAE (false click beetles)

Subfamily Eucneminae

"*Deltometopus*" *fossilis* Wickham 1916: UCM

"*Eucnemis*" *antiquatus* Wickham 1914: MCZ

Subfamily Macraulacinae

"*Fornax*" *relictus* Wickham 1916: USNM

Subfamily Melasinae

"*Microrhagus*" *miocenicus* Wickham 1916: UCM

"*Microrhagus*" *vulcanicus* Wickham 1916: MCZ

Family LYCIDAE (net-winged beetles)

†*Miocaenia pectinicornis* Wickham 1914: MCZ

Family LAMPYRIDAE (lightning bugs and fireflies)

Subfamily Lampyrinae

"*Lucidota*" *prima* (Wickham) Wickham 1914: UCM

Family CANTHARIDAE (soldier beetles)

Subfamily Cantharinae

"*Cantharis*" *hesperus* (Wickham) Wickham 1920: MCZ

"*Cantharis*" *humatus* (Wickham) Wickham 1920: USNM

"*Podabrus*" *cupesoides* Wickham 1917: USNM

"*Podabrus*" *florissantensis* Wickham 1914: MCZ

"*Podabrus*" *fragmentatus* Wickham 1914: MCZ

"*Podabrus*" *wheeleri* Wickham 1909: YPM

Subfamily Silinae

"*Polemius*" *crassicornis* Wickham 1914: MCZ

Subfamily Chauliognathinae

"*Chauliognathus*" *pristinus* Scudder 1876: USNM

"*Trypherus*" *aboriginalis* Wickham 1913: USNM

Family DERMESTIDAE (dermestid or skin beetles)

Subfamily Dermestinae

"*Dermestes*" *tertiarius* Wickham 1912: AMNH

Subfamily Orphilinae
    "*Orphilus*" *dubius* Wickham 1912: UCM
Subfamily Attageninae
    "*Attagenus*" *aboriginalis* Wickham 1913: USNM
    "*Attagenus*" *sopitus* Scudder 1900: MCZ
Family BOSTRICHIDAE (branch and twig borers)
    †*Protapate contorta* Wickham 1912: YPM
Subfamily Bostrichinae
    "*Amphicerus*" *sublaevis* Wickham 1914: MCZ
    "*Xylobiops*" *lacustre* Wickham 1912: AMNH
Subfamily Dinoderinae
    "*Dinoderus*" *cuneicollis* Wickham 1913: USNM
Family ANOBIIDAE (anobiid beetles)
    †*Gastrallanobium subconfusum* Wickham 1914: USNM
Subfamily Ernobiinae
    "*Ernobius*" *effectus* Wickham 1914: MCZ
    "*Xestobium*" *alutaceum* Wickham 1913: USNM
Subfamily Anobiinae
    "*Anobium*" *durescens* Scudder 1900: MCZ
    "*Oligomerus*" *brevisculus* Wickham 1916: USNM
    "*Oligomerus*" *druatus* Wickham 1914: MCZ
    "*Oligomerus: florissantensis* Wickham 1914: MCZ
Subfamily Xyletininae
    "*Vrilletta*" *monstrosa* Wickham 1917: USNM
    "*Vrilletta*" *tenuistriata* Wickham 1913: USNM
Family LYMEXYLIDAE (shiptimber beetles)
Subfamily Melittommatinae
    "*Melittomma*" *lacustrinum* (Wickham) Wickham 1920: UCM
Family TROGOSSITIDAE (bark-gnawing beetles)
Subfamily Peltinae
    "*Ostoma*" *laminata* (Wickham) Wickham 1920: YPM
Subfamily Trogossitinae
    "*Tenebroides*" *corrugata* Wickham 1913: USNM
Family CLERIDAE (checkered beetles)
    †*Miohydnocera wolcotti* (Wickham) Mawdsley 1992: USNM

Subfamily Clerinae
    Thanasimus florissantensis (Wickham) Mawdsley 1992: MCZ
Subfamily Epiphloeinae
    Epiphloeus pristinus (Wickham) Mawdsley 1992: MCZ
Subfamily Korynetinae
    "*Necrobia*" *divinatoria* Wickham 1914: MCZ
    "*Necrobia*" *sibylla* Wickham 1914: USNM
Family MELYRIDAE (soft-winged flower beetles)
Subfamily Dasytinae
    †"*Eudasytites*" *listriformis* Wickham 1912: AMNH
    "*Trichochrous*" *miocenus* Wickham 1912: UCM
Subfamily Malachiinae
    "*Collops*" *desuetus* Wickham 1914: MCZ
    "*Collops*" *extrusus* Wickham 1914: MCZ
    "*Collops*" *priscus* Wickham 1914: MCZ
    "*Malachius*" *immurus* Wickham 1917: USNM
Family BRACHYPTERIDAE
    "*Amartus*" *petrefactus* Wickham 1912: UCM
Family NITIDULIDAE (sap beetles)
    †*Cychramites hirtus* Wickham 1913: USNM
    †*Epanuraea ingenita* Scudder 1900: MCZ
    †*Miophenolia cilipes* Wickham 1916: USNM
Subfamily Carpophilinae
    "*Carpophilus*" *restructus* Scudder 1900: MCZ
Subfamily Nitidulinae
    "*Nitidula*" *prior* Scudder 1900: MCZ
Subfamily Cillaeinae
    "*Colopterus*" *pygidialis* [= *Colastus pygidialis* Wickham 1913]: USNM
Family CUCUJIDAE (flatbark beetles)
    †*Lithocoryne arcuata* Wickham 1913: USNM
    †*Lithocoryne coloradensis* Wickham 1914: USNM
    †*Lithocoryne gravis* Scudder 1900: MCZ
Subfamily Cucujinae
    "*Pediacus*" *periclitans* Scudder 1900: MCZ
Family CRYPTOPHAGIDAE (silken fungus beetles)
Subfamily Cryptophaginae

"*Antherophagus*" *megalops* Wickham 1913: USNM

"*Cryptophagus*" *bassleri* Wickham 1913: USNM

"*Cryptophagus*" *petricola* Wickham 1916: USNM

"*Cryptophagus*" *scudderi* Wickham 1914: MCZ

Family EROTYLIDAE (pleasing fungus beetles)

    Subfamily Tritominae

        "*Triplax*" *diluviana* (Wickham) 1920: MCZ

        "*Triplax*" *materna* (Wickham) Wickham 1920: UCM

        "*Triplax*" *petrefacta* (Wickham) Wickham 1920: USNM

        "*Triplax*" *submersa* (Wickham) Wickham 1920: UCM

Family COCCINELLIDAE (ladybird beetles)

    Subfamily Chilocorinae

        "*Chilocorus*" *ulkei* Scudder 1900: MCZ

    Subfamily Coccinellinae

        "*Adalia*" *subversa* Scudder 1900: MCZ

        "*Anatis*" *resurgens* Wickham 1917: USNM

        "*Coccinella*" *florissantensis* Wickham 1914: MCZ

        "*Coccinella*" *sodoma* Wickham 1913: USNM

Family LATRIDIIDAE (minute brown scavenging beetles)

    Subfamily Corticariina

        "*Corticaria*" *aeterna* Wickham 1914: USNM

        "*Corticaria*" *egregia* Wickham 1914: MCZ

        "*Corticaria*" *occlusa* Wickham 1914: MCZ

        "*Corticaria*" *petrefacta* Wickham 1913: USNM

Family MYCETOPHAGIDAE (hairy fungus beetles)

    "*Mycetophagus*" *exterminatus* Wickham 1913: USNM

    "*Mycetophagus*" *willistoni* Wickham 1913: USNM

Family MORDELLIDAE (tumbling flower beetles)

    Subfamily Mordellinae

        "*Mordella*" *lapidicola* Wickham 1909: NHM, YPM

        "*Mordella*" *stygia* Wickham 1914: MCZ

        "*Mordellistena*" *florissantensis* Wickham 1912: UCM

        "*Mordellistena*" *nearctica* Wickham 1914: USNM

        "*Mordellistena*" *protogaea* Wickham 1914: USNM

        "*Mordellistena*" *scudderiana* Wickham 1914: USNM

        "*Mordellistena*" *smithiana* Wickham 1913: USNM

        "*Tomoxia*" *inundata* Wickham 1914: MCZ

Family RHIPIPHORIDAE (wedge-shaped beetles)

    Subfamily Rhipiphorinae

        "*Macrosiagon*" *geikiei* Scudder 1890: MCZ

Family COLYDIIDAE (cylindrical bark beetles)

    Subfamily Colydiinae

        †*Phloeonemites miocenus* Wickham 1912: UCM

        "*Eucicones*" *oblongopunctata* Wickham 1913: USNM

        †*Rhagoderidea striata* Wickham 1914: MCZ

Family TENEBRIONIDAE (darkling beetles)

    "*Miostenosis*" *lacordairei* Wickham 1913: USNM

    "*Pactostoma*" *prima* (Wickham) Wickham 1920: YPM

    †*Proteleates centralis* Wickham 1914: USNM

    †*Protoplatycera laticornis* Wickham 1914: MCZ

    †*Tenebrionites alatus* Cockerell 1927: NHM

    Subfamily Tenebrioninae

        "*Blapstinus*" *linellii* Wickham 1913: USNM

        "*Ephalus*"(?) *adumbratus* Scudder 1900: MCZ

        "*Meracantha*" *lacustris* Wickham 1909: YPM

        "*Ulus*" *minutus* Wickham 1914: USNM

    Subfamily Alleculinae

        "*Capnochroa*" *senilis* Wickham 1913: USNM

        Hymenorus haydeni Wickham 1914: MCZ

        "*Isomira*" *aurora* Wickham 1914: [specimen has not been located]

        "*Isomira*" *florissantensis* Wickham 1914: MCZ

        "*Pseudocistela*" *antiqua* Wickham 1913: USNM

        "*Pseudocistela*" *vulcanica* Wickham 1914: MCZ

    Subfamily Diaperinae

        "*Platydema*" *antiquorum* Wickham 1912: UCM

        "*Platydema*" *bethunei* Wickham 1913: USNM

Family SYNCHROIDAE (false darkling beetles)

    "*Synchroa*" *quiescent* Wickham 1911: UCM

Family OEDEMERIDAE (false blister beetles)
  †*Paloedemera crassipes* Wickham 1914: MCZ
  Subfamily Oedemerinae
    "*Copidita*" *miocenica* Wickham 1914: MCZ
Family MELOIDAE (blister beetles)
  "*Epicauta*" *subneglecta* Wickham 1914: MCZ
  "*Gnathium*" *aetatis* Scudder 1900: MCZ
  "*Nemognatha*" *exsecta* Wickham 1912: UCM
  "*Tetraonyx*" *minuscula* Wickham 1914: MCZ
Family PYROCHROIDAE (fire-colored beetles)
  Subfamily Pedilinae
    Pedilus calypso [= *Corphyra calypso* Wickham 1914]: MCZ
Family SALPINGIDAE (narrow-waisted bark beetles)
  †*Pythoceropsis singularis* Wickham 1913: USNM
Family ANTHICIDAE (antlike leaf beetles)
  Subfamily Macratriinae
    †*Lithomacratria mirabilis* Wickham 1914: MCZ
    Macratria gigantea Wickham 1910: YPM
Family CERAMBYCIDAE (long-horned beetles)
  Subfamily Spondylidinae
    Arhopalus(?) *pavitus* (Cockerell) Linsley 1942: NHM
    †*Protospondylis florissantensis* (Wickham) Linsley: USNM
  Subfamily Lepturinae
    Anoplodera antecurrens (Wickham) Linsley 1942: USNM
    Grammoptera nanella (Wickham) Linsley 1942: MCZ
    "*Gaurotes*" *striatopunctatus* Wickham 1914: MCZ
    "*Leptura*" *petrorum* Wickham 1912: UCM
    "*Leptura*" *ponderosissima* Wickham 1913: USNM
    "*Leptura*" *wickhami* Linsley 1942: MCZ
    Pidonia ingenua (Wickham) Linsley 1942: MCZ
    Pidonia leidyi (Wickham) Linsley 1942: USNM
  Subfamily Cerambycinae
    Anelaphus extinctus (Wickham) Linsley 1942: MCZ
    †*Callidiopsites grandiceps* Wickham 1913: UCM
    "*Callimoxys*" *primordialis* Wickham 1911: UCM
    Cyllene florissantensis (Wickham) Linsley 1942: MCZ
    "*Dryobius*" *miocenicus* Beutenmüller and Cockerell 1908: AMNH, UCM
    "*Elaphidion*" *fracticorne* Wickham 1911: UCM
    †*Palaeosmodicum hamiltoni* Wickham 1914: USNM
    "*Phymatodes*" *grandaevum* (Wickham) Linsley 1942: USNM
    "*Phymatodes*" *miocenicus* Wickham 1914: MCZ
    "*Phymatodes*" *volans* Beutenmüller and Cockerell 1908: AMNH
    Semanotus puncticollis (Wickham) Linsley 1942: USNM
    "*Stenosphenus*" *pristinus* Wickham 1914: MCZ
  Subfamily Lamiinae
    "*Leptostylus*" *scudderi* Wickham 1914: MCZ
    Parolamia rudis Scudder 1878: MCZ
    Psapharochus lengii (Wickham) Linsley 1942: USNM
    †*Protipochus vandykei* Wickham 1914: MCZ
    †*Protoncideres primus* Wickham 1913: USNM
    Saperda florissantensis Wickham 1916: USNM
    Saperda lesquereuxi Cockerell 1916: UCM
    Saperda submersa Cockerell 1908: UCM
Family CHRYSOMELIDAE (leaf beetles)
  †*Crioceridea dubia* Wickham 1912: UCM
  "*Luperodes*" *submonilis* Wickham 1914: USNM
  †*Oryctoscirtetes protogaeum* Scudder 1876: MCZ
  †*Plectrotetrophanes hageni* Wickham 1914: MCZ
  †*Prochaetocnema florissantella* Wickham 1914: MCZ
  Subfamily Bruchinae
    "*Bruchus*" *aboriginalis* Wickham 1914: MCZ
    "*Bruchus*" *antaeus* Wickham 1917: USNM
    "*Bruchus*" *bowditchi* Wickham 1912: UCM
    "*Bruchus*" *carpophiloides* Wickham 1914: MCZ
    "*Bruchus*" *dormescens* Wickham 1913: USNM
    "*Bruchus*" *exhumatus* Wickham 1912: UCM

"*Bruchus*" *florissantensis* Wickham 1912: USNM
"*Bruchus*" *haywardi* Wickham 1912: UCM
"*Bruchus*" *henshawi* Wickham 1912: AMNH
"*Bruchus*" *osborni* Wickham 1912: UCM
"*Bruchus*" *primoticus* Wickham 1914: MCZ
"*Bruchus*" *scudderi* Wickham 1912: UCM
"*Bruchus*" *submersus* Wickham 1914: MCZ
"*Bruchus*" *succintus* Wickham 1913: USNM
"*Bruchus*" *wilsoni* Wickham 1913: USNM
"*Spermophagus*" *pluto* Wickham 1914: MCZ
"*Spermophagus*" *vivifacatus* Scudder 1876: MCZ

Subfamily Donaciinae
"*Donacia*" *primaeva* Wickham 1912: AMNH

Subfamily Criocerinae
"*Lema*" *evanescens* Wickham 1910: YPM
"*Lema*" *fortior* Wickham 1914: MCZ
"*Lema*" *lesquereuxi* Wickham 1914: USNM

Subfamily Hispinae
Odontota americana Wickham 1914: MCZ

Subfamily Chrysomelinae
Chrysomela vesperalis Scudder 1900: MCZ

Subfamily Galerucinae
"*Altica*" *renovata* [= "*Haltica*" *renovata* Wickham 1914]: MCZ
"*Diabrotica*" *bowditchiana* Wickham 1914: MCZ
"*Diabrotica*" *exesa* Wickham 1911: AMNH
"*Diabrotica*" *florissantella* Wickham 1914: MCZ
"*Diabrotica*" *uteana* Wickham 1914: MCZ
"*Systena*" *florissantensis* Wickham 1913: USNM
"*Trirhabda*" *majuscula* Wickham 1914: MCZ
"*Trirhabda*" *megacephala* Wickham 1914: MCZ
"*Trirhabda*" *sepulta* Wickham 1914: MCZ

Subfamily Cryptocephalinae
"*Cryptocephalus*" *miocenus* Wickham 1913: USNM
Colaspis aetatis Wickham 1911: UCM
Colaspis diluvialis Wickham 1914: MCZ
Colaspis luti Scudder 1900: MCZ
Colaspis proserpina Scudder 1914: MCZ
"*Metachroma*" *florissantensis* Wickham 1912: UCM

Family ANTHRIBIDAE (fungus weevils)
"*Brachytarsus*" *dubius* Wickham 1913: USNM
"*Euparius*" *adumbratus* [= *Cratoparis adumbrates* Wickham 1911]: UCM
"*Euparius*" *arcessitus* [= *Cratoparis arcessitus* Scudder 1893]: MCZ
†*Saperdirhynchus priscotitillator* Scudder 1893: MCZ
†*Stiraderes conradi* Scudder 1893: MCZ

Subfamily Anthribinae
"*Anthribus*" *sordidus* Scudder 1893: MCZ
"*Tropideres*" *vastatus* Scudder 1893: MCZ

Family BRENTIDAE (straight-snouted beetles)
Subfamily Apioninae
"*Apion*" *cockerelli* Wickham 1911: AMNH
"*Apion*" *confectum* Scudder 1893: MCZ
"*Apion*" *curiosum* Scudder 1893: MCZ
"*Apion*" *exanimale* Scudder 1893: MCZ
"*Apion*" *florissantensis* Wickham 1916: USNM
"*Apion*" *pumilum* Scudder 1893: MCZ
"*Apion*" *refrenatum* Scudder 1893: MCZ
"*Apion*" *scudderianum* Wickham 1916: USNM
"*Apion*" *smithii* Scudder 1893: MCZ

Family ATTELABIDAE (leaf-rolling beetles)
†*Docirhynchus culex* Scudder 1893: MCZ
†*Docirhynchus ibis* Wickham 1912: UCM
†*Docirhynchus terebrans* Scudder 1893: MCZ
†*Isothea alleni* Scudder 1893: MCZ
†*Masteutes rupis* Scudder 1893: MCZ
†*Masteutes saxifer* Scudder 1893: MCZ
†*Paltorhynchus narwhal* Scudder 1893: MCZ, USNM
†*Paltorhynchus rectirostris* Scudder 1893: MCZ
"*Rhynchites*" *laminarum* Wickham 1916: USNM
"*Rhynchites*" *subterraneus* Scudder 1893: MCZ
"*Rhynchites*" *vulcan* Wickham 1916: USNM
†*Teretrum primulum* Scudder 1893: MCZ
†*Toxorhynchus grandis* Wickham 1911: UCM
†*Toxorhynchus minusculus* Scudder 1893: MCZ

†*Toxorhynchus oculatus* Scudder 1893: MCZ

†*Trypanorhynchus corruptivus* Scudder 1893: MCZ

†*Trypanorhynchus depratus* Scudder 1893: MCZ, USNM

†*Trypanorhynchus exilis* Wickham 1913: USNM

†*Trypanorhynchus minutissimus* Wickham 1913: USNM

†*Trypanorhynchus obliquus* Wickham 1913: USNM

†*Trypanorhynchus sedatus* Scudder 1893: MCZ

Family CURCULIONIDAE (weevils or snout beetles)

†*Catobaris coenosa* Scudder 1893: MCZ

†*Centron moricollis* Scudder 1893: MCZ

Coeliodes primotinus Scudder 1893: MCZ

"*Coniatus*" *differens* Wickham 1912: USNM

"*Coniatus*" *evisceratus* Scudder 1893: MCZ

"*Cremastorhynchus*" *stabilis* Scudder 1893: MCZ, USNM

"*Cyphus*" *florissantensis* Wickham 1914: MCZ

"*Cyphus*" *subterraneus* Wickham 1911: UCM

†*Eocleonus subjectus* Scudder 1893: MCZ

"*Erirhinus*" *dormitus* Scudder 1893: MCZ

†*Eugnamptidea robusta* Wickham 1916: USNM

†*Eugnamptidea tertiaria* Wickham 1912: AMNH

†*Eudomus pinguis* Scudder 1893: MCZ, USNM

†*Eudomus robustus* Scudder 1893: MCZ, USNM

†*Evopes occubatus* Scudder 1893: MCZ, USNM

†*Evopes veneratus* Scudder 1893: MCZ

†*Geralophus antiquarius* Scudder 1893: MCZ

†*Geralophus discessus* Scudder 1893: MCZ

†*Geralophus fossicius* Scudder 1893: MCZ, USNM

†*Geralophus lassatus* Scudder 1893: MCZ, USNM

†*Geralophus occultus* Scudder 1893: [specimen has not been located]

†*Geralophus pumiceus* Scudder 1893: MCZ

†*Geralophus repositus* Scudder 1893: MCZ

†*Geralophus retritus* Scudder 1893: MCZ

†*Geralophus saxuosus* Scudder 1893: MCZ

†*Geralophus scudderi* Wickham 1911: UCM

†*Hipporhinops sternbergi* Cockerell 1926: NHM

†*Laccopygus nilesii* Scudder 1893: MCZ

†*Lithophthorus rugosicollis* Scudder 1893: MCZ

†*Miogeraeus recurrens* Wickham 1916: USNM

†*Numitor claviger* Scudder 1893: MCZ

†*Oligocryptus sectus* (Scudder) Carpenter 1985: MCZ

†*Oryctorhinus tenuirostris* Scudder 1893: MCZ

"*Scyphophorus*" *fossionis* Scudder 1893: MCZ

"*Scyphophorus*" *laevis* Scudder 1893: MCZ

"*Scyphophorus*" *tertiarius* Wickham 1911: AMNH

"*Sitona*" *exitiorum* Scudder 1893: MCZ

†*Tenillus firmus* Scudder 1893: USNM

Subfamily Brachycerinae

"*Hormorus*" *saxorum* Scudder 1893: MCZ

"*Menoetius*" *humatus* [= "*Lachnopus*" *humatus* Scudder 1893]: MCZ

"*Menoetius*" *recuperatus* [= "*Lachnopus*" *recuperates* Scudder 1893]: MCZ

"*Omileus*" *evanidus* Scudder 1893: MCZ

"*Ophryastes*" *championi* Wickham 1912: USNM

†*Ophryastites absconsus* Scudder 1893: MCZ

†*Ophryastites cinereus* Scudder 1893: USNM

†*Ophryastites miocenus* Wickham 1912: USNM

†*Otiorhynchites absentivus* Scudder 1893: [specimen has not been located]

†*Otiorhynchites contusa* Cockerell [publication has not been located]: USNM

†*Otiorhynchites florissantensis* Wickham 1911: UCM

"*Pandeleteius*" *nudus* Wickham 1917: USNM

†*Rhysosternum aeternabile* Scudder 1893: MCZ

†*Rhysosternum longirostre* Scudder 1893: MCZ

†*Smicrorhynchus macgeei* Scudder 1893: MCZ

†*Spodotribus terrulentus* Scudder 1893: MCZ

"*Trigonoscuta*" *inventa* Scudder 1893: MCZ

Subfamily Curculioninae

"*Acalles*" *exhumatus* Wickham 1913: USNM

"*Acalyptus*" *obtusus* Scudder 1893: MCZ

"*Anthonomus*" *arctus* Scudder 1893: MCZ

"*Anthonomus*" *concussus* Scudder 1893: MCZ

"*Anthonomus*" *corruptus* Scudder 1893: MCZ

"*Anthonomus*" *debilatus* Scudder 1893: MCZ

"*Anthonomus*" *defossus* Scudder 1876: MCZ

"*Anthonomus*" *evigilatus* Scudder 1893: MCZ

"*Anthonomus*" *primordius* Scudder 1893: MCZ

"*Anthonomus*" *principalis* (Scudder) Wickham 1920: MCZ

"*Anthonomus*" *requiescens* (Scudder) Wickham 1920: MCZ

"*Anthonomus*" *reventus* Scudder 1893: MCZ

"*Anthonomus*" *rohweri* Wickham 1912: USNM

"*Auleutes*" *florissantensis* Wickham 1913: USNM

"*Auleutes*" *wymani* Scudder 1893: MCZ

"*Aulobaris*" *damnata* Scudder 1893: USNM

"*Baris*" *antediluviana* Wickham 1916: USNM

"*Baris*" *cremastorhynchoides* Wickham 1913: USNM

"*Baris*" *divisa* Scudder 1893: MCZ

"*Baris*" *florissantensis* Wickham 1913: USNM

"*Baris*" *harlani* Scudder 1893: MCZ

"*Baris*" *hoveyi* Wickham 1912: AMNH

"*Baris*" *imperfecta* Scudder 1893: MCZ

"*Baris*" *matura* Scudder 1893: MCZ

"*Baris*" *nearctica* Wickham 1916: USNM

"*Baris*" *primalis* Wickham 1917: USNM

"*Baris*" *renovata* Wickham 1916: USNM

"*Baris*" *schucherti* Wickham 1912: USNM

"*Centrinus*" *hypogaeus* Wickham 1916: USNM

"*Centrinus*" *obnuptus* Scudder 1893: MCZ

"*Centrinus*" *vulcanicus* Wickham 1913: USNM

"*Ceutorhynchus*" *blaisdelli* Wickham 1916: USNM

"*Ceutorhynchus*" *clausus* Scudder 1893: MCZ

"*Ceutorhynchus*" *compactus* Scudder 1893: MCZ

"*Ceutorhynchus*" *duratus* Scudder 1893: MCZ

"*Ceuthorhynchus*" *evinctus* Scudder 1893: [specimen has not been located]

"*Cleonis*" *degeneratus* Scudder 1893: MCZ

"*Cleonis*" *estriatus* Wickham 1912: AMNH

"*Cleonis*" *exterraneus* Scudder 1893: MCZ

"*Cleonis*" *foersteri* Scudder 1893: USNM

"*Cleonis*" *primoris* Scudder 1893: USNM

"*Cleonis*" *rohweri* Wickham 1911: UCM

Coccotorus principalis Scudder 1893: MCZ

Coccotorus requiescens Scudder 1893: MCZ

"*Conotrachelus*" *florissantensis* Wickham 1912: USNM

"*Cryptorhynchus*" *coloradensis* Wickham 1912: USNM

"*Cryptorhynchus*" *fallii* Wickham 1912: USNM

"*Cryptorhynchus*" *kerri* Scudder 1893: MCZ

"*Cryptorhynchus*" *profusus* Scudder 1893: MCZ

"*Curculio*" *anicularis* [= "*Balaninus*" *anicularis* Scudder 1893]: MCZ

"*Curculio*" *duttoni* [= "*Balaninus*" *duttoni* Scudder 1893]: MCZ

"*Curculio*" *extinctus* [= "*Balaninus*" *extinctus* Wickham 1912]: USNM

"*Curculio*" *femoratus* [= "*Balaninus*" *femoratus* Scudder 1893]: MCZ, USNM

"*Curculio*" *flexirostris* [= "*Balaninus*" *flexirostris* Scudder 1893]: MCZ

"*Curculio*" *florissantensis* [= "*Balaninus*" *florissantensis* Wickham 1913]: USNM

"*Curculio*" *minusculoides* [= "*Balaninus*" *minusculoides* Wickham 1911]: UCM

"*Curculio*" *minusculus* [= "*Balaninus*" *minusculus* Scudder 1893]: MCZ

"*Curculio*" *restrictus* [= "*Balaninus*" *restrictus* Scudder 1893]: MCZ

"*Dorytomus*" *coercitus* Scudder 1893: MCZ

"*Dorytomus*" *williamsi* Scudder 1893: MCZ

"*Grypus*" *curvirostris* [= "*Grypidius*" *curvirostris* Scudder 1893]: MCZ

"*Gymnetron*" *antecurrens* Scudder 1893: MCZ

"*Hylobius*" *lacoei* Scudder 1893: USNM

"*Macrorhoptus*" *intutus* Scudder 1893: MCZ

"*Magdalis*" *sedimentorum* Scudder 1893: MCZ

"*Magdalis*" *striaticeps* Wickham 1911: UCM

"Notaris" brevicollis [= *Erycus brevicollis* Scudder 1893]: MCZ

"Odontopus" irvingii [= "*Prionomerus*" irvingii Scudder 1893]: MCZ

"Pachybaris" rudis Wickham 1912: UCM

"Procas" verberatus Scudder 1893: MCZ

"Rynchaenus" languidulus [= "*Orchestes*" languidulus Scudder 1893]: MCZ

"Tychius" evolatus Scudder 1893: MCZ, USNM

"Tychius" ferox Wickham 1917: USNM

"Tychius" secretus Scudder 1893: MCZ

"Tychius" whitneyi (Scudder) Wickham 1920: MCZ

Subfamily Cossoninae

"Cossonus" gabbii Scudder 1893: MCZ

Family SCOLYTIDAE (bark beetles, engraver beetles, and ambrosia beetles)

†*Adipocephalus hydropicus* Wickham 1916: USNM

†*Pityophthoridea diluvialis* Wickham 1916: USNM

†*Xyleborites longipennis* Wickham 1913: USNM

Subfamily Hylesininae

"Hylastes" americanus Wickham 1913: USNM

"Hylesinus" extractus Scudder 1893: AMNH

"Hylurgops" piger Wickham 1913: USNM

"Phthorophloeus" zimmermanni (Wickham) Wickham 1920: USNM

ORDER MECOPTERA (scorpionflies and hangingflies)

Family PANORPIDAE (common scorpionflies)

†*Holcorpa maculosa* Scudder 1878: MCZ

Panorpa arctiiformis Cockerell 1907: UCM

Panorpa rigida Scudder 1890: MCZ

†Family EOMEROPIDAE

†*Eomerope tortriciformis* Cockerell 1909: YPM

ORDER DIPTERA (flies)

Family TIPULIDAE (crane flies)

†*Cladoneura willistoni* Scudder 1894: [specimen has not been located]

†*Cyttaromyia cancellata* Scudder 1894: [specimen has not been located]

†*Cyttaromyia clathrata* Scudder 1894: [specimen has not been located]

†*Cyttaromyia oligocena* Scudder 1894: [specimen has not been located]

†*Cyttaromyia princetoniana* Scudder 1894: USNM

†*Limnocema lutescens* Scudder 1894: MCZ

†*Limnocema marcescens* Scudder 1894: MCZ

†*Limnocema mortoni* Scudder 1894: MCZ

†*Limnocema sternbergi* Cockerell 1927: NHM

†*Limnocema styx* Scudder 1894: MCZ

Limnophila rogersii Scudder 1894: [specimen has not been located]

Limnophila ruinarum Scudder 1894: [specimen has not been located]

Limnophila strigosa Scudder 1894: [specimen has not been located]

Limnophila vasta Scudder 1894: [specimen has not been located]

†*Manapsis anomala* Scudder 1894: MCZ

†*Oryctogma sackenii* Scudder 1894: MCZ

†*Rhadinobrochus extinctus* Scudder 1894: MCZ

Rhabdomastix praecursor Cockerell 1927: NHM

Rhamphidia faecaria Scudder 1894: [specimen has not been located]

Rhamphidia loewi Scudder 1894: [specimen has not been located]

Rhamphidia saxetana Scudder 1894: [specimen has not been located]

Subfamily Limoniinae

Antocha principialis Scudder 1894: MCZ

Cladura integra Scudder 1894: MCZ

Cladura maculata Scudder 1894: MCZ

Dicranomyia fontaini Scudder 1894: [specimen has not been located]

Dicranomyia fragilis Scudder 1894: [specimen has not been located]

Dicranomyia inferna Scudder 1894: [specimen has not been located]

Dicranomyia longipes Scudder 1894: [specimen has not been located]

Dicranomyia stagnorum Scudder 1894: USNM

Gonomyia frigida Scudder 1894: MCZ

Gonomyia labefactata Scudder 1894: MCZ

Gonomyia primogenitalis Scudder 1894: MCZ, USNM

Gonomyia profundi Scudder 1894: MCZ

Subfamily Tipulinae

Tipula bilineata [= *Tipulidea bilineata* Scudder 1894]: MCZ

Tipula carolinae Scudder 1894: MCZ

Tipula clauda Scudder 1894: MCZ, USNM

Tipula consumpta [= *Tipulidea consumpta* Scudder 1894]: MCZ

Tipula evanitura Scudder 1894: MCZ, USNM

Tipula florissanta Scudder 1894: MCZ, USNM

Tipula heilprini Scudder 1894: MCZ

Tipula hepialina Cockerell 1909: UCM

Tipula internecata Scudder 1894: MCZ

Tipula lapillescens Scudder 1894: MCZ

Tipula lethaea Scudder 1894: MCZ

Tipula limi Scudder 1894: MCZ, USNM

Tipula maclurei Scudder 1894: MCZ

Tipula magnifica Scudder 1894: MCZ

Tipula needhami Cockerell 1910: UCM

Tipula paludis [= *Micrapsis paludis* Scudder 1894]: MCZ

Tipula picta [= *Tipulidea picta* Scudder 1894]: MCZ

Tipula reliquiae [= *Tipulidea reliquiae* Scudder 1894]: MCZ

Tipula revivificata Scudder 1894: MCZ

Tipula rigens Scudder 1894: MCZ

Tipula subterjacens Scudder 1894: MCZ

Tipula tartari Scudder 1894: MCZ

Tipula wilmattae Melander 1949: AMNH

Family BIBIONIDAE (march flies)

Subfamily Bibioninae

Bibio atavus Cockerell 1909: AMNH, NHM, UCM

Bibio capnodes Melander 1949: USNM

Bibio cockerelli James 1937: UCM

Bibio excurvatus Melander 1949: USNM

Bibio explanatus Melander 1949: AMNH

Bibio jamesi Melander 1949: AMNH

Bibio podager Melander 1949: AMNH

Bibio vetus James 1937: UCM

Bibio vulcanius Melander 1949: USNM

Bibio wickhami Cockerell 1914: USNM

Bibiodes intermedia James 1937: UCM

Subfamily Hesperinae

Hesperinus immutabilis Melander 1949: USNM

Subfamily Pleciinae

Penthetria longifurca Melander 1949: USNM

Plecia axeliana Cockerell 1914: UCM, USNM

Plecia decapitata Cockerell 1917: USNM

Plecia explanata Cockerell 1917: USNM

Plecia gradata Melander 1949: AMNH

Plecia melanderi Cockerell 1911: AMNH

Plecia orycta Melander 1949: USNM

Plecia tessella Melander 1949: USNM

Family MYCETOPHILIDAE (fungus gnats)

†*Proapemon infernus* Melander 1949: AMNH

Subfamily Mycetophilinae

Exechia priscula Melander 1949: AMNH

†*Mycetophaetus intermedius* Scudder 1892: MCZ

Mycetophila bradenae Cockerell 1915: UCM

Subfamily Sciophilinae

Boletina hypogaea Melander 1949: USNM

Lasiosoma mirandula Cockerell 1909: UCM

Leia miocenica Cockerell 1911: UCM

Mycomya cockerelli Johannsen 1912: YPM

Mycomya lithomendax Cockerell 1914: MCZ

Family SCIARIDAE (dark-winged fungus gnats, root gnats)

Sciara dormitans Melander 1949: AMNH

Sciara florissantensis Cockerell 1916: USNM

Sciara requieta Melander 1949: AMNH

Sciara sopora Melander 1949: AMNH

Family CECIDOMYIIDAE (gall midges or gall gnats)
  Subfamily Cecidomyiidae
    Cecidomyia pontaniiformis Cockerell 1908: NHM
Family SCATOPSIDAE (minute black scavenger flies)
  Subfamily Scatopsinae
    Reichertella fasciata Melander 1949: USNM
Family PTYCHOPTERIDAE (phantom crane flies)
  Subfamily Bittacomorphinae
    Bittacomorpha miocenica Cockerell 1910: UCM
Family CHIRONOMIDAE (midges)
  Subfamily Chironominae
    Chironomus pausatus Melander 1949: AMNH
    Chironomus primaevus Melander 1949: AMNH
    Chironomus pristinus Melander 1949: AMNH
    Chironomus proterus Melander 1949: NHM
    Chironomus requiescens Melander 1949: AMNH
    Chironomus scudderiellus Cockerell 1916: USNM
    Chironomus sepultus Melander 1949: AMNH
  Subfamily Diamesinae
    Diamesa extincta Melander 1949: AMNH
Family TABANIDAE (horse and deer flies)
  Subfamily Crysopinae
    Silvius merychippi (Cockerell) Melander 1946: USNM
  Subfamily Tabaninae
    Tabanus hipparionis Cockerell 1909: AMNH
    Tabanus parahippi Cockerell 1909: AMNH
Family RHAGIONIDAE (snipe flies)
    Atrichops hesperius Cockerell 1914: UCM
    Leptis mystaceaeformis Cockerell 1909: AMNH, UCM
    Rhagio fossitus Melander 1949: USNM
    Rhagio wheeleri Melander 1949: AMNH
    Symphoromyia subtrita Cockerell 1911: AMNH
Family XYLOPHAGIDAE (xylophagid flies)
  Subfamily Coenomyiinae
    Dialysis revelata Cockerell 1908: UCM
Family XYLOMYIDAE (xylomyid flies)
    Solva inornata Melander 1949: USNM
    Solva moratula [= *Xylomyia moratula* Cockerell 1914]: UCM
Family STRATIOMYIDAE (soldier flies)
    Lasiopa carpenteri James 1937: MCZ
    †*Moyamyia limigena* Melander 1949: AMNH
    Rhingiopsis prisculus (Cockerell) James 1937: AMNH
  Subfamily Beridinae
    Beris miocenica James 1937: UCM
  Subfamily Clittelarinae
    Cyphomyia rohweri (Cockerell) James 1937: UCM
  Subfamily Stratiomylinae
    Oxycera contusa Cockerell 1917: USNM
Family THEREVIDAE (stiletto flies)
  Subfamily Therevinae
    Nebritus willistoni Melander 1949: USNM
    Rueppellia vagabunda Cockerell 1927: NHM
    Psilocephala hypogaea Cockerell 1909: AMNH
    Psilocephala scudderi Cockerell 1909: AMNH
Family MYDAIDAE (mydas flies)
  Subfamily Mydainae
    Mydas miocenicus Cockerell 1913: UCM
Family ASILIDAE (robber flies and grass flies)
    Senobasis borealis James 1939: MCZ
    Senoprosopis antiquus James 1939: UMMP
    Senoprosopis eureka Melander 1949: AMNH
    Senoprosopis romeri Hull 1957: MCZ
  Subfamily Asilinae
    Asilus amelanchieris Cockerell 1911: UCM
    Asilus circulionis Melander 1946: USNM
    Asilus florissantinus James 1939: UCM
    Asilus peritulus Cockerell 1909: UCM
    Asilus wickhami Cockerell 1914: USNM, UCM
    Philonicus saxorum James 1939: UCM
  Subfamily Dasypogoninae
    Ceraturgus praecursor [= *Ceraturgopsis praecursor* James 1939]: UCM
    Cophura antiquella Cockerell 1913: UCM
    Dioctria florissantina (Cockerell) Cockerell 1909: UCM

Dioctria pulveris Cockerell 1917: USNM

Lestomyia miocenica James 1939: UCM

Microstylum(?) *destructum* Cockerell 1909: NHM, UCM

Microstylum wheeleri Cockerell 1908: UCM

Saropogon oblitescens Cockerell 1914: UCM

Taracticus contusus Cockerell 1910: UCM

Taracticus renovatus Cockerell 1911: UCM

Nicocles miocenicus Cockerell 1909: UCM

Subfamily Leptogastrinae

Leptogaster prior Melander 1946: USNM

Family NEMESTRINIDAE (tangle-veined flies)

Prosoeca florigerus [= *Palembolus florigerus* Scudder 1878]: MCZ

Subfamily Himoneurinae

Hirmoneura willistoni (Cockerell) Bequaert & Carpenter 1936: AMNH

Subfamily Nemestrininae

Neorhynchocephalus(?) *melanderi* (Cockerell) Bequaert & Carpenter 1936: YPM

Neorhynchocephalus occultator (Cockerell) Bequaert & Carpenter 1936: UCM, NHM

Neorhynchocephalus vulcanicus (Cockerell) Bequaert & Carpenter 1936: YPM

Family BOMBYLIIDAE (bee flies)

†*Acreotrichites scopulicornis* Cockerell 1917: USNM

†*Alepidophora cockerelli* Melander 1949: AMNH

†*Alepidophora minor* Melander 1949: AMNH

†*Alepidophora pealei* Cockerell 1909: YPM

†*Alomatia fusca* Cockerell 1914: USNM

†*Lithocosmus coquilletti* Cockerell 1909: AMNH, NHM

†*Lithocosmus palpalis* (Cockerell) Evenhuis 1984: USNM

†*Megacosmus mirandus* Cockerell 1909: AMNH

†*Megacosmus secundus* Cockerell 1911: UCM

†*Melanderella glossalis* Cockerell 1909: AMNH

†*Melanderella testea* Melander 1949: USNM

†*Pachysystropus condemnatus* Cockerell 1910: UCM

†*Pachysystropus rohweri* Cockerell 1909: YPM

†*Protepacmus setosus* Cockerell 1916: USNM

†*Protolomatia antiqua* Cockerell 1914: USNM

†*Protolomatia recurrens* Cockerell 1916: USNM

†*Tithonomyia atra* (Melander) Evenhuis 1984: USNM

†*Verrallites cladurus* Cockerell 1913: UCM

Subfamily Gerontinae

Geron platysoma Cockerell 1914: MCZ

†*Geronites stigmalis* Cockerell 1914: UCM

Subfamily Systropodinae

Dolichomyia tertiaria Cockerell 1917: USNM

Subfamily Tomomyzinae

Amphicosmus delicatulus Melander 1949: AMNH

Subfamily Usiinae

Apolysis magister Melander 1947: AMNH

Family EMPIDIDAE (dance flies)

Acallomyia probolaea Melander 1949: AMNH

†*Progloma rohweri* James 1937: UCM

Subfamily Empidinae

Empis florissantana Cockerell 1914: UCM, USNM

Empis infossa Melander 1949: USNM

Empis miocenica Cockerell 1914: MCZ

Empis perdita Cockerell 1916: UCM

Rhamphomyia aeterna Melander 1949: AMNH

Rhamphomyia calvimontis Cockerell 1916: [specimen has not been located]

Rhamphomyia craterae Melander 1949: USNM

Rhamphomyia fossa Melander 1949: USNM

Rhamphomyia hypolitha Cockerell 1917: USNM

Rhamphomyia inanimata Melander 1949: USNM

Rhamphomyia infernalis Melander 1949: USNM

Rhamphomyia interita Melander 1949: AMNH

Rhamphomyia morticina Melander 1949: NHM

Rhamphomyia senecta Melander 1949: AMNH

Rhamphomyia sepulta Cockerell 1916: UCM

Rhamphomyia spodites Melander 1949: AMNH

Rhamphomyia tumulata Melander 1949: AMNH

Subfamily Tachydromiinae

    Tachypeza primitiva Melander 1949: USNM

Family PLATYPEZIDAE (flat-footed flies)

    †*Eucallimyia fortis* Cockerell 1911: AMNH, UCM

Family PHORIDAE (hump-backed flies)

    Paraspiniphora laminarum (Brues) Cockerell 1913: AMNH

  Subfamily Phorinae

    Phora cockerelli Brues 1908: AMNH

    Phora tumbae Melander 1949: AMNH

Family SYRPHIDAE (hover, flower, or syrphid flies)

    †*Cacogaster novamaculata* Hull 1945: MCZ

    †*Protochrysotoxum sphinx* Hull 1945: AMNH

  Subfamily Eustalinae

    Cheilosia hecate Hull 1945: MCZ

    Cheilosia miocenica Cockerell 1909: AMNH, UCM

    Cheilosia sepultula Cockerell 1916: USNM

    Chrysogaster antiquaria Hull 1945: MCZ

    Pipiza melanderi Hull 1945: AMNH

    Rhingia zephyrea Hull 1945: MCZ

    Sphegina obscura Hull 1945: MCZ

  Subfamily Syrphinae

    †*Archalia femorata* Hull 1945: MCZ, USNM

    Leucozona nigra Hull 1945: AMNH

    Platycheirus lethaeus Melander 1949: UCM

    Syrphus aphidopsidis Cockerell 1909: NHM, UCM

    Syrphus carpenteri Hull 1945: MCZ

    Syrphus platychiralis Hull 1945: MCZ

    Syrphus willistoni Cockerell 1909: AMNH

Family PIPUNCULIDAE (big-headed flies)

    Protonephrocerus florissantius Carpenter & Hull 1939: MCZ

Family OTITIDAE (picture-winged flies)

  Subfamily Otitinae

    Melieria atavina Cockerell 1917: USNM

    Melieria calligrapha Melander 1949: AMNH

Family RICHARDIIDAE (richardiid flies)

    †*Pachysomites inermis* Cockerell 1916: USNM

    †*Urortalis caudatus* Cockerell 1917: USNM

Family PIOPHILIDAE (skipper flies)

  Subfamily Piophilinae

    Mycetaulus incretus Melander 1949: AMNH

Family AGROMYZIDAE (leaf-miner flies)

  Subfamily Agromyzinae

    Agromyza praecursor Melander 1949: AMNH

    Melanagromyza prisca Melander 1949: AMNH

    Melanagromyza tephrias Melander 1949: AMNH

Family SCIOMYZIDAE (marsh flies)

  Subfamily Sciomyzinae

    Sciomyza florissantensis Cockerell 1909: AMNH, UCM

Family SEPSIDAE (black scavenger flies)

  Subfamily Sepsinae

    Themira saxifica Melander 1949: AMNH

Family LAUXANIIDAE (lauxaniid flies)

    Sapromyza veterana Melander 1949: USNM

Family HELEOMYZIDAE (heleomyzid flies)

    Heteromyiella miocenica Cockerell 1914: UCM

Family SCATHOPHAGIDAE (dung flies)

    Cordylura exhumata Cockerell 1916: USNM

Family ANTHOMYIIDAE (anthomyiid flies)

  Subfamily Anthomiinae

    Anthomyia atavella Cockerell 1913: UCM

    Anthomyia persepulta Cockerell 1917: USNM

    Mecistoneuron perpetuum Melander 1949: AMNH

    Ophyra vetusta Melander 1949: AMNH

Family MUSCIDAE (muscid flies)

  Subfamily Glossininae (tsetse flies)

    Glossina oligocenus (Scudder) Cockerell 1907: MCZ

    Glossina osborni Cockerell 1909: NHM

ORDER TRICHOPTERA (caddisflies)

Family HYDROPSYCHIDAE (net-spinning caddisflies)

  Subfamily Hydropsychinae

Hydropsyche marcens Scudder 1890: MCZ

Hydropsyche scudderi Cockerell 1909: NHM

†*Leptobrochus luteus* Scudder 1890: MCZ

†*Litobrochus externatus* Scudder 1890: MCZ

†*Mesobrochus imbecillus* Scudder 1890: MCZ

†*Mesobrochus lethaeus* Scudder 1890: MCZ

Paladicella eruptionis Scudder 1890: MCZ

Family POLYCENTROPIDAE (trumpet-net and tube-making caddisflies)

†*Derobrochus abstractus* Scudder 1890: MCZ

†*Derobrochus caenulentus* Scudder 1890: MCZ

†*Derobrochus commoratus* Scudder 1890: MCZ

†*Derobrochus craterae* Scudder 1890: MCZ, USNM

†*Derobrochus frigescens* Scudder 1890: MCZ

†*Derobrochus marcidus* Scudder 1890: MCZ

†*Derobrochus typharum* Cockerell 1910: UCM

Polycentropus aeternus (Scudder) Cockerell 1907: MCZ

Polycentropus eviratus Scudder 1890: MCZ

Polycentropus exesus Scudder 1890: MCZ

Family PSYCHOMYIIDAE (tube-making and trumpet-net caddisflies)

Tinodes paludigena Scudder 1890: MCZ

Family LIMNEPHILIDAE (northern caddisflies)

†*Tricheopteryx florissantensis* (Cockerell) Cockerell 1927: AMNH, UCM

Subfamily Limnephilinae

Limnephilus soporatus Scudder 1890: MCZ

Family PHRYGANEIDAE (large caddisflies)

†*Limnopsyche dispersa* Scudder 1890: MCZ

Neuronia evanescens Scudder 1890: MCZ

Phryganea labefacta Scudder 1890: MCZ

Phryganea miocenica Cockerell 1913: UCM

Phryganea wickhami Cockerell 1914: USNM

Family ODONTOCERIDAE

†*Phenacopsyche larvalis* Cockerell 1927: NHM

†*Phenacopsyche vexans* Cockerell 1909: NHM, UCM

Family LEPTOCERIDAE (long-horned caddisflies)

Setodes abbreviata Scudder 1890: [specimen has not been located]

Setodes portionalis Scudder 1890: MCZ

Family incertae sedis

Indusia cypridis Cockerell 1910: AMNH

ORDER LEPIDOPTERA (butterflies, moths, and skippers)

Family LYONETIIDAE (lyonetiid moths)

Bucculatrix sp. Opler 1982: [specimen has not been located]

Family OECOPHORIDAE (oecophorid moths)

Subfamily Ethmiinae

Ethmia mortuella [= *Psecadia mortuella* Scudder 1890]: MCZ

Family COSSIDAE (carpenter and leopard moths)

†*Adelopsyche frustrans* Cockerell 1926: UCM

Family TORTRICIDAE (tortricid moths)

Subfamily Tortricinae

Tortrix(?) *destructus* Cockerell 1916: USNM

Tortrix florissantana Cockerell 1907: NHM, UCM

Family CASTNIIDAE (butterfly moths)

†*Dominickus castnioides* Tindale 1985: FMNH

Family PIERIDAE (white and sulphur butterflies)

Subfamily Pierinae

†*Oligodonta florissantensis* Brown 1976: WC, UF

†*Stolopsyche libytheoides* Scudder 1889: MCZ

Family LIBYTHEIDAE (snout butterflies)

Subfamily Libytheinae

†*Barbarothea florissanti* Scudder 1892: [specimen has not been located]

†*Prolibythea vagabunda* Scudder 1889: MCZ

Family NYMPHALIDAE (brush-footed butterflies)

Subfamily Nymphalinae

†*Apanthesis leuce* Scudder 1886: MCZ

Doxocopa wilmattae [= *Chlorippe wilmattae* Cockerell 1907]: MCZ

Jupitellia charon (Scudder) Carpenter 1985: USNM

†*Lithopsyche styx* Scudder 1889: MCZ

†*Nymphalites obscurum* Scudder 1889: MCZ

†*Nymphalites scudderi* Beutenmüller & Cockerell 1908: AMNH, UCM

†*Prodryas persephone* Scudder 1878: MCZ

Vanessa amerindica Miller & Brown 1989: UF

Family GEOMETRIDAE (measuring worms, geometers, and cankerworms)

Subfamily Larentiinae

Hydriomena(?) *protrita* Cockerell 1922: AMNH

Family SATURNIIDAE (royal moths and giant silkworm moths)

Attacus(?) *fossilis* Cockerell 1912: UCM

Family incertae sedis

†*Phylledestes vorax* Cockerell 1907: UCM

ORDER HYMENOPTERA (sawflies, parasitic wasps, ants, wasps, and bees)

Family BLASTICOTOMIDAE

Runaria ostenta (Brues) Zhelochovtzev & Rasnitsyn 1972: MCZ

Family XYELIDAE (sawflies)

Subfamily Macroxyelinae

Megaxyela petrefacta Brues 1908: MCZ

Family PAMPHILIIDAE (leaf-rolling and web-spinning sawflies)

†*Atocus defessus* Scudder 1892: MCZ

Subfamily Pamphiliinae

Neurotoma cockerelli Rohwer 1908: AMNH, UCM

Family ARGIDAE

Subfamily Sterictiphorinae

Sterictiphora konowi (Rohwer) Zhelochovtzev & Rasnitsyn 1972: UCM

Family CIMBICIDAE

Subfamily Cimbicinae

†*Trichiosomites obliviosus* Brues 1908: MCZ

Subfamily Phenacoperginae

†*Phenacoperga coloradensis* (Cockerell) Cockerell 1908: NHM, UCM

†*Pseudocimbex clavatus* Rohwer 1908: UCM

Family TENTHREDINIDAE (common sawflies)

Dineura cockerelli Rohwer 1908: AMNH

Dineura fuscipennis Rohwer 1908: UCM

Dineura laminarum Brues 1908: MCZ

Dineura microsoma Cockerell 1927: NHM

Dineura saxorum Cockerell 1906: AMNH

Pteronus prodigus Brues 1908: MCZ

†*Taeniurites fortis* Cockerell 1917: USNM

Tenthredella fenestralis Cockerell 1927: NHM

Tenthredella oblita Cockerell 1917: USNM

Tenthredella toddi Cockerell 1914: UCM

Subfamily Phyllotominae

Eriocampoides micrarche Cockerell 1916: USNM

Eriocampoides minus Cockerell 1917: USNM

Eriocampoides revelatus Cockerell 1909: NHM, UCM

Subfamily Allantinae

Athalia wheeleri (Cockerell) Zhelochovtzev & Rasnitsyn 1972: AMNH

Eriocampa bruesi Rohwer 1908: UCM

Eriocampa celata Cockerell 1914: [specimen has not been located]

Eriocampa disjecta Cockerell 1922: DMNH

Eriocampa pristina Cockerell 1910: AMNH

Eriocampa scudderi Brues 1908: MCZ

Eriocampa synthetica Cockerell 1911: UCM

Eriocampa wheeleri Cockerell 1906: AMNH

†*Palaeotaxonus trivittatus* Rohwer 1908: AMNH

†*Palaeotaxonus typicus* Brues 1908: MCZ

†*Palaeotaxonus vetus* Cockerell 1917: USNM

Pseudosiobla megoura Cockerell 1907: [specimen has not been located]

Subfamily Heterarthrinae

Fenusa parva (Brues) Zhelochovtzev & Rasnitsyn 1972: AMNH

Fenusa primula Rohwer 1908: AMNH

Subfamily Nemainae

Cladius petrinus Cockerell 1914: UCM, USNM

†*Eohemichroa eophila* (Cockerell) Zhelochovtzev & Rasnitsyn 1972: AMNH

†*Florissantinus angulatus* Zhelochovtzev & Rasnitsyn 1972: MCZ

Hoplocampa ilicis Cockerell 1927: NHM

Subfamily Selandriinae

Selandria sapindi Cockerell 1910: UCM

Subfamily Tenthredininae

Macrophya pervetusta Brues 1908: MCZ

Mesoneura vexabilis (Brues) Zhelochovtzev & Rasnitsyn 1972: MCZ

†*Nortonella typica* Rohwer 1908: UCM

Tenthredo avia Brues 1908: MCZ

Tenthredo infossa Brues 1908: MCZ

Tenthredo misera Brues 1908: MCZ

Tenthredo saxorum Rohwer 1908: UCM

Tenthredo submersa Cockerell 1907: AMNH, UCM

Family CEPHIDAE (stem sawflies)

Subfamily Cephinae

Janus disperditus Cockerell 1913: USNM, UCM

Family SIRICIDAE (horntails)

†*Lithoserix williamsi* Brown 1986: UCM

Family AULACIDAE

Subfamily Aulacinae

Aulacostethus secundus [= *Aulacites secundus* Cockerell 1916]: USNM

Aulacus bradleyi Brues 1910: UCM, MPM

Pristaulacus rohweri Brues 1910: AMNH

Family BRACONIDAE (braconids or parasitic wasps)

Subfamily Alysiinae

Alysia exigua Brues 1910: MCZ

Alysia petrina Brues 1910: MCZ

Alysia phanerognatha Cockerell 1927: NHM

Alysia ruskii Cockerell 1913: UCM

Subfamily Agathidinae

Agathis juvenilis Brues 1910: AMNH

Agathis saxatilis Brues 1910: MCZ

Agathis velatus Brues 1910: MCZ

Subfamily Blacinae

Calyptus wilmattae Brues 1910: AMNH

Subfamily Braconinae

Bracon abstractus Brues 1910: MCZ

Bracon cockerelli Brues 1910: AMNH

Bracon florissanticola Cockerell 1919: [specimen has not been located]

Bracon resurrectus Brues 1910: AMNH

Subfamily Cheloninae

Chelonus depressus Brues 1910: MCZ

Chelonus muratus Brues 1910: MCZ

Chelonus solidus Brues 1910: AMNH

Subfamily Euphorinae

Euphorus indurescens Brues 1910: MCZ

Subfamily Helconinae

Diospilus repertus Brues 1910: MCZ

Diospilus soporatus (Brues) Mason 1976: MCZ

Urosigalphus aeternus Brues 1910: AMNH

Subfamily Ichneutinae

†*Oligoneuroides destructus* Brues 1910: MCZ

Subfamily Microgasterinae

Microgaster primordialis Brues 1906: AMNH

Microplitis vesperus Brues 1910: AMNH

Subfamily Rogadinae

Exothecus abrogatus Brues 1910: AMNH

Rogus tertiarius Brues 1906: AMNH

Family ICHNEUMONIDAE (ichneumonids)

Absyrtus decrepitus Brues 1910: MCZ

Acoenites defunctus Brues 1906: AMNH

Amblyteles pealei Cockerell 1927: NHM

Camperotops solidatus Brues 1910: MCZ

Exochilum inusitatum Brues 1910: MCZ

Hellwigia obsoleta (Brues) Townes 1966: MCZ

†*Hiatensor funditus* Brues 1910: MCZ

†*Hiatensor semirutus* Brues 1910: MCZ

Labrorychus latens Brues 1910: MCZ

Lapton daemon Brues 1910: MCZ

Leptobatopsis ashmeadii Brues 1910: UCM

Limnerium consuetum Brues 1910: AMNH
Limnerium depositum Brues 1910: AMNH
Limnerium plenum Brues 1910: AMNH, UCM
Limnerium tectum Brues 1910: MCZ
Limnerium vetustum Brues 1910: MCZ
Parabates memorialis Brues 1910: UCM
Tylecomnus davisii Brues 1910: MCZ
Tylecomnus pimploides Brues 1910: MCZ
Xylonomus sejugatus Brues 1910: MCZ

Subfamily Anomaloninae
   Anomalon confertum Brues 1910: MCZ
   Anomalon deletum Brues 1910: MCZ
   Anomalon excisum Brues 1910: MCZ
   Anomalon miocenicum Cockerell 1919: MCZ, UCM
   Barylypa primigenia Brues 1910: MCZ
   Demophorus antiquus Brues 1910: MCZ

Subfamily Banchinae
   Exetastes inveteratus Brues 1910: MCZ

Subfamily Campopleginae
   Porizon exsectus Brues 1910: MCZ

Subfamily Cryptinae
   Cryptus delineatus Brues 1910: MCZ
   Hemiteles lapidescens Brues 1910: MCZ
   Hemiteles obtectus Brues 1910: MCZ
   Hemiteles priscus Brues 1910: MCZ
   Hemiteles veternus Brues 1910: AMNH, UCM
   Mesoleptus apertus Brues 1910: MCZ
   Mesoleptus exstirpatus Brues 1910: AMNH
   Mesostenus modestus Brues 1906: AMNH

Subfamily Ichneumoninae
   Ichneumon alpha Brues 1910: MCZ
   Ichneumon cannoni Brues 1910: [specimen has not been located]
   Ichneumon concretus Brues 1910: MCZ
   Ichneumon decrepitus Brues 1910: MCZ
   Ichneumon dormitans Brues 1910: MCZ
   Ichneumon exesus Brues 1910: MCZ
   Ichneumon obduratus Brues 1910: MCZ
   Ichneumon pollens Brues 1910: MCZ
   Ichneumon primigenius Brues 1910: MCZ
   Ichneumon provectus Brues 1910: MCZ
   Ichneumon somniatus Brues 1910: MCZ
   Ichneumon torpefactus Brues 1910: MCZ
   Trogus vetus Brues 1910: MCZ

Subfamily Mesochorinae
   Mesochorus abolitus Brues 1910: AMNH
   Mesochorus aboriginalis Brues 1910: MCZ
   Mesochorus carceratus Brues 1910: AMNH
   Mesochorus cataclysmi Brues 1910: MCZ
   Mesochorus dormitorius Brues 1910: MCZ
   Mesochorus lapideus Brues 1910: MCZ
   Mesochorus revocatus Brues 1910: MCZ
   Mesochorus terrosus Brues 1910: MCZ

Subfamily Metopiinae
   Exochus captus Brues 1910: AMNH

Subfamily Orthocentrinae
   Orthocentrus defossus Brues 1910: MCZ
   Orthocentrus primus Brues 1906: AMNH

Subfamily Pimplinae
   Glypta aurora Brues 1910: MCZ
   Lampronota pristina Brues 1910: MCZ
   Lampronota tenebrosa Brues 1910: UCM, AMNH
   Lampronota stygialis Brues 1910: MCZ
   †*Megatryphon mortiferus* Cockerell 1924: [specimen has not been located]
   †*Mesopimpla sequoiarum* Cockerell 1919: USNM
   Pimpla appendigera Brues 1906: AMNH
   Pimpla morticina Brues 1910: MCZ
   Pimpla rediviva Brues 1910: MCZ
   Pimpla revelata Brues 1910: MCZ
   Pimpla senilis Brues 1910: MCZ
   Polysphincta inundata Brues 1910: MCZ
   Polysphincta mortuaria Brues 1910: MCZ
   Polysphincta petrorum Brues 1910: MCZ
   Theronia wickhami Cockerell 1919: MCZ, UCM

Tryphon cadaver Brues 1910: MCZ
Tryphon explanatum Cockerell 1919: NHM, UCM
Tryphon florissantensis Brues 1910: AMNH
Tryphon lapideus Brues 1910: MCZ
Tryphon peregrinus Brues 1910: MCZ
Tryphon senex Brues 1910: AMNH, UCM
Subfamily Rhyssinae
Rhyssa petiolata Brues 1906: AMNH
Family TORYMIDAE (torymids)
Subfamily Toryminae
†*Palaeotorymus aciculatus* Brues 1910: MCZ
†*Palaeotorymus laevis* Brues 1910: MCZ
†*Palaeotorymus striatus* Brues 1910: MCZ
†*Palaeotorymus typicus* Brues 1910: MCZ
Torymus sackeni Brues 1910: MCZ
Family AGAONIDAE (fig wasps)
Tetrapus mayri Brues 1910: MCZ
Family PTEROMALIDAE (pteromalids)
Ormyrodes petrefactus Brues 1910: MCZ
†*Bruesisca submersus* (Brues) Heiquist 1961: MCZ
Subfamily Pteromalinae
Pteromalus exanimis Brues 1910: AMNH, UCM
Family EURYTOMIDAE (seed chalcidoids)
Subfamily Eurytominae
Eurytoma sepulta Brues 1910: AMNH
Eurytoma sequax Brues 1910: AMNH
Family CHALCIDIDAE (chalcids)
Subfamily Chalcidinae
Chalcis perdita Brues 1910: MCZ
Chalcis praevalens Cockerell 1907: AMNH, NHM
Chalcis tortilis Brues 1910: MCZ
Eterochalcis scudderi (Brues) Burks 1939: MCZ
Family STEPHANIDAE (stephanids)
Subfamily Stephaninae
†*Protostephanus ashmeadi* Cockerell 1906: MCZ
Family CYNIPIDAE (gall wasps)
Subfamily Cynipinae
Aulacidea ampliforma Kinsey 1919: MCZ

Aulacidea progenitrix Kinsey 1919: MCZ
Andricus myrica Brues 1910: MCZ
Family IBALIIDAE (ibaliids)
Subfamily Ibaliinae
†*Protoibalia connexiva* Brues 1910: MCZ
Family FIGITIDAE (pupal parasites)
Subfamily Figitinae
Figites solus Brues 1910: AMNH
Family PROCTOTRUPIDAE
Proctotrupes exhumatus Brues 1910: MCZ
Family DIAPRIIDAE (diapriids)
†*Galesimorpha wheeleri* Brues 1910: AMNH
Subfamily Belytinae
Belyta mortuella Brues 1906: AMNH
Pantoclis deperdita Brues 1906: AMNH
Subfamily Diapriinae
Paramesius defectus Brues 1910: MCZ
Family SCELIONIDAE (scelionids)
†*Palaeoteleia oxyura* Cockerell 1914: MCZ
Family CHRYSIDIDAE (cuckoo wasps)
Subfamily Chrysidinae
Chrysis miocenica Rohwer 1909: UCM
Chrysis rohweri Cockerell 1907: AMNH, UCM
Family BETHYLIDAE (bethylids)
Subfamily Bethyinae
Epyris deletus Brues 1910: AMNH
Family SPHECIDAE (sphecid wasps)
†*Prophilanthus destructus* Cockerell 1906: MCZ
Subfamily Sphecinae
Ammophila antiquella Cockerell 1906: MCZ
Subfamily Sceliphroninae
Chalybion mortuum Cockerell 1907: UCM
Subfamily Crabroninae
Tracheliodes mortuellus Cockerell 1906: MCZ
Crabro longaevus Cockerell 1910: UCM
Pison cockerellae Rohwer 1908: AMNH
Subfamily Nyssoninae
†*Hoplisidia kohliana* Cockerell 1906: MCZ

Hoplisus sepultus Cockerell 1906: MCZ

Mellinus handlirschi Brues 1908: UCM

Subfamily Pemphredoninae

Passaloecus fasciatus Rohwer 1909: UCM

Passaloecus scudderi Cockerell 1906: MCZ

Subfamily Philanthinae

Philanthus saxigenus Rohwer 1909: UCM

Family MELITTIDAE (melittids)

Subfamily Melittinae

Melitta willardi Cockerell 1909: UCM

Family HALICTIDAE (sweat bees)

†*Cyrtapis anomalus* Cockerell 1908: UCM

Subfamily Halictinae

Halictus florissantellus Cockerell 1906: MCZ

Halictus miocenicus Cockerell 1909: UCM

Halictus scudderiellus Cockerell 1906: MCZ

Family ANDRENIDAE (andrenid bees)

Subfamily Andreninae

Andrena(?) *clavula* Cockerell 1906: MCZ

Andrena grandipes Cockerell 1911: UCM

Andrena hypolitha Cockerell 1908: UCM

Andrena percontusa Cockerell 1914: MCZ

Andrena sepulta Cockerell 1906: MCZ

†*Lithandrena saxorum* Cockerell 1906: MCZ

Subfamily Panurginae

†*Libellulapis antiquorum* Cockerell 1906: MCZ

†*Libellulapis wilmattae* Cockerell 1913: UCM, USNM

†*Pelandrena reducta* Cockerell 1909: UCM

Family MEGACHILIDAE (leaf-cutting bees)

Subfamily Megachilinae

Anthidium exhumatum Cockerell 1906: MCZ, UCM

Anthidium scudderi Cockerell 1906: MCZ

Dianthidium tertiarium Cockerell 1906: MCZ

Heriades bowditchi Cockerell 1906: MCZ

Heriades halictinus Cockerell 1906: MCZ

Heriades laminarum Cockerell 1906: MCZ

Heriades mersatus Cockerell 1923: UCM

Heriades mildredi Cockerell 1925: UCM

Heriades priscus Cockerell 1917: USNM

Heriades saxuosus Cockerell 1913: UCM

†*Lithanthidium pertriste* Cockerell 1911: UCM

Megachile praedicta Cockerell 1908: UCM

Stelis seneciophilia Cockerell 1908: [specimen has not been located]

Family ANTHOPHORIDAE (cuckoo bees, digger bees, and carpenter bees)

†*Protomelecta brevipennis* Cockerell 1908: UCM

Subfamily Anthophorinae

Anthophora melfordi Cockerell 1908: NHM, UCM

Subfamily Xylocopinae

Ceratina disrupta Cockerell 1906: MCZ

Family APIDAE (bumble bees, honey bees, and orchid bees)

Subfamily Bombinae (bumble bees)

Bombus florissantensis (Cockerell) Zeuner & Manning 1976: MCZ

Family TIPHIIDAE (tiphiids)

Subfamily Anthoboscinae

Anthobosca foxiana [= *Geotiphia foxiana* Cockerell 1906]: MCZ

Anthobosca halictina [= *Geotiphia halictina* Cockerell 1910]: UCM

Anthobosca pachysoma [= *Geotiphia pachysoma* Cockerell 1927]: NHM

Anthobosca sternbergi [= *Geotiphia sternbergi* Cockerell 1910]: AMNH

Subfamily Tiphiinae

Paratiphia praefracta Cockerell 1907: UCM

†*Lithotiphia scudderi* Cockerell 1906: MCZ

Family POMPILIDAE (spider-hunting wasps)

Subfamily Pepsinae

Cryptocheilus hypogaeus Cockerell 1914: USNM

Pepsis avitula Cockerell 1941: [specimen has not been located]

Subfamily Ceropalinae

Agenia cockerellae Rohwer 1909: UCM

Agenia saxigena Cockerell 1908: YPM, NHM

†*Ceropalites infelix* Cockerell 1906: MCZ

Salius florissantensis (Cockerell) Rohwer 1909: MCZ

Salius laminarum Rohwer 1909: UCM

Salius scudderi (Cockerell) Rohwer 1909: UCM, MCZ

Salius senex Rohwer 1909: UCM, NHM

Family SCOLIIDAE

†*Floriscolia relicta* Rasnitsyn 1993: AMNH

Family VESPIDAE (paper wasps, yellow jackets, hornets, mason wasps, and potter wasps)

Subfamily Eumeninae

Odynerus palaeophilus Cockerell 1906: MCZ

Odynerus percontusus Cockerell 1914: MCZ, UCM

Odynerus praesepultus Cockerell 1906: MCZ

Odynerus terryi Cockerell 1909: UCM

Odynerus wilmattae Cockerell 1914: UCM

Subfamily Vespinae

†*Palaeovespa florissantia* Cockerell 1906: MCZ

†*Palaeovespa gillettei* Cockerell 1906: MCZ

†*Palaeovespa relecta* Cockerell 1923: [specimen has not been located]

†*Palaeovespa scudderi* Cockerell 1906: MCZ

†*Palaeovespa wilsoni* Cockerell 1906: MCZ, UCM

Family FORMICIDAE (ants)

Subfamily Dolichoderinae

Dolichoderus antiquus Carpenter 1930: MCZ

Dolichoderus rohweri Carpenter 1930: MCZ

†*Elaeomyrmex coloradensis* Carpenter 1930: MCZ

†*Elaeomyrmex gracilis* Carpenter 1930: MCZ

Iridomyrmex florissantius Carpenter 1930: MCZ

Iridomyrmex obscurans Carpenter 1930: USNM

Liometopum miocenicum Carpenter 1930: MCZ

Liometopum scudderi Carpenter 1930: MCZ

†*Mianeuretus mirabilis* Carpenter 1930: MCZ

Microtus miocenicus Cockerell 1927: NHM

†*Miomyrmex impactus* Carpenter 1930: NHM

†*Miomyrmex striatus* Carpenter 1930: MCZ

†*Petraeomyrmex minimus* Carpenter 1930: MCZ

†*Protazteca capitata* Carpenter 1930: MCZ

†*Protazteca elongata* Carpenter 1930: MCZ

†*Protazteca quadrata* Carpenter 1930: MCZ

Subfamily Formicinae

Camponotus fuscipennis Carpenter 1930: MCZ

Camponotus microcephalus Carpenter 1930: MCZ

Camponotus petrifactus Carpenter 1930: MCZ

Formica cockerelli Carpenter 1930: MCZ

Formica grandis Carpenter 1930: USNM

Formica robusta Carpenter 1930: MCZ

Subfamily Myrmicinae

Aphaenogaster donisthorpei Carpenter 1930: MCZ

Aphaenogaster mayri Carpenter 1930: MCZ

†*Archimyrmex rostratus* Cockerell 1923: USNM

†*Cephalomyrmex rotundatus* Carpenter 1930: MCZ

†*Eulithomyrmex rugosus* (Carpenter) Carpenter 1935: MCZ

†*Eulithomyrmex striatus* (Carpenter) Carpenter 1935: MCZ

Pheidole tertiaria Carpenter 1930: MCZ

Pogonomyrmex fossilis Carpenter 1930: MCZ

Tetramorium peritulum Cockerell 1927: NHM

Subfamily Ponerinae

†*Archiponera wheeleri* Carpenter 1930: MCZ

Subfamily Pseudomyrminae

Pseudomyrmex extincta Carpenter 1930: MCZ

OTHER INVERTEBRATES

Phylum ARTHROPODA

Class Crustacea

Subclass Ostracoda (ostracods)

ORDER PODOCOPIDA

Family CYPRIDIDAE

Cypris florissantensis Cockerell 1910: AMNH

Phylum MOLLUSCA

Class Gastropoda (snails)

Subclass Pulmonata

ORDER LYMNOPHILA (lung-bearing freshwater snails)

Family LYMNAEIDAE

Lymnaea florissantica Cockerell 1908: UCM

Lymnaea scudderi Cockerell 1906: AMNH

Lymnaea sieverti Cockerell 1906: AMNH

Family PLANORBIDAE

Planorbis florissantensis Cockerell 1906: UCM

ORDER GEOPHILA (land snails)

Family VITRINIDAE

Omphalina(?) *laminarum* Cockerell 1906: [specimen has not been located]

Vitrea fagalis Cockerell 1907: UCM

Class Bivalvia (clams)

ORDER VENEROIDEA

Family PISIDIIDAE

Sphaerium florissantense Cockerell 1906: AMNH

VERTEBRATES

Class Pisces (fishes)

ORDER AMIIFORMES

Family AMIIDAE (bowfins)

Amia scutata Cope 1875: USNM

ORDER CYPRINIFORMES

Family CATOSTOMIDAE (suckers)

†*Amyzon commune* Cope 1874: USNM

†*Amyzon fusiforme* Cope 1875: [specimen has not been located]

†*Amyzon pandatum* Cope 1875: USNM

ORDER SILURIFORMES (catfishes)

Family ICTALURIDAE

Ictalurus pectinatus (Cope) Lundberg 1975: USNM

ORDER PERCOPSIFORMES

Family APHREDODERIDAE (pirate perches)

†*Trichophanes foliarum* Cope 1878: USNM

Class Aves (birds)

ORDER CHARADRIIFORMES

Family unknown

Undescribed shorebird

ORDER CUCULIFORMES (cuckoos)

Family CUCULIDAE

†*Eocuculus cherpinae* Chandler 1999: DMNH

ORDER CORACIIFORMES (African rollers and their relatives)

Family unknown

†*Palaeospiza bella* Allen 1878 [reclassified by Olson 1985]: MCZ

Incertae sedis (affinities unknown)

Charadrius sheppardianus Cope 1881: AMNH

†*Yalavis tenuipes* Shufeldt 1913: YPM (questionable whether actually from Florissant)

Feather [= *Fontinalis pristina* Lesquereux 1883, described as a moss]: USNM

"Passerine bird" Shufeldt 1917

Class Mammalia (mammals)

ORDER MARSUPIALIA

Family DIDELPHIDAE

†*Herpetotherium* cf. *huntii* [= *Peratherium* cf. *huntii* (Cope), of Gazin 1935]: USNM (non-type)

ORDER PERISSODACTYLA

Family EQUIDAE

†*Mesohippus* sp.

†Family BRONTOTHERIIDAE
    Genus indeterminate (brontothere)
†Family cf. HYRACODONTIDAE
    cf. †*Hyracodon* (small rhinoceros)

ORDER ARTIODACTYLA
†Family MERYCOIDODONTIDAE [= OREODONTIDAE]
    †*Merycoidodon* sp. (oreodont)
†Family LEPTOMERYCIDAE?
    †*Leptomeryx*?

APPENDIX 2

Museums with Significant Florissant Collections

Many museums and universities are known to house significant collections of Florissant fossils. Most of the museums on the list below have type or otherwise published specimens. In addition, other museums and universities (only a few of which are listed here) have small unpublished collections from Florissant. In total, at least 40,000 Florissant specimens are housed at these museums. Almost all of these collections are for research, and only about 300 specimens—mostly at Florissant Fossil Beds National Monument and the Denver Museum of Nature & Science—are on public exhibit.

The counts for number of type specimens combines holotypes, syntypes, paratypes, and others, and are based on original descriptions of species, some of which were synonymized in subsequent literature. Customized searches for such information can be accomplished on the Florissant database website, but researchers who would like to use these museum data are required to contact the individual museums for permission.

AMNH
American Museum of Natural History
New York, New York
This is a moderate-sized collection of fossil insects and spiders and also includes a few fish and plants. The estimated total collection size is about 1,000 specimens, about 300 of which have been cited in publications, including 194 types. Most of the collection came from Cockerell's Florissant expeditions of 1906–1908.

CAS
California Academy of Sciences
San Francisco, California
This is a small collection that includes one insect holotype.

CM
Carnegie Museum of Natural History
Pittsburgh, Pennsylvania
This is a small collection of about 100 specimens of plants and insects including nine paratypes, along with one cast of the only known oreodont specimen.

CSM
Colorado School of Mines
Golden, Colorado
This is a small collection of about 120 specimens.

DMNH
Denver Museum of Nature & Science (formerly Denver Museum of Natural History)
Denver, Colorado
This is a large collection of more than 3,200 specimens, mostly of plants but also many insects as well as the only cuckoo from Florissant. The museum has a small Florissant exhibit in the Prehistoric Journey hall.

FLFO
Florissant Fossil Beds National Monument
Florissant, Colorado
This is a large collection of plants, insects, and fish, with more than 2,700 cataloged specimens as of 2002. The monument's public exhibit consists of about 220 specimens, most of which are on loan from Waynesburg College.

FMNH
Field Museum of Natural History
Chicago, Illinois
This is a moderate-sized collection consisting of more than 1,200 specimens of plants, insects, and fish, with 21 published specimens including three types.

GLAHM
Hunterian Museum, University of Glasgow
Glasgow, Scotland
This is a small collection of plants, including two that are on public display.

MCZ
Museum of Comparative Zoology, Harvard University
Cambridge, Massachusetts
This is the second largest Florissant collection, containing an estimated 8,000 specimens, most of which are fossil insects and spiders. There are 2,290 published specimens including 2,071 types. This includes 90 percent of Samuel Scudder's published specimens, making it one of the most significant collections from Florissant. In addition, there is a small collection of fossil plants and one specimen of a bird.

MPM
Milwaukee Public Museum
Milwaukee, Wisconsin
This is a moderate-sized collection of about 1,000 specimens, mostly of fossil plants but also including insects. There are 19 published specimens including 17 that are indicated as types.

MSU
Michigan State University
East Lansing, Michigan
This collection consists of microscopic slides of pollen and spores.

NHM
The Natural History Museum, London (formerly British Museum of Natural History)
London, England
This is a moderate-sized and high-quality collection consisting of more than 1,100 specimens of plants, insects, spiders, and fish. It contains 127 published specimens including 85 types. Many of these were collected by the Cockerell expeditions of 1906–1908 and are counterparts of specimens that are at the University of Colorado or elsewhere.

NMS
National Museums of Scotland
Edinburgh, Scotland
This is a small collection of 18 fossil plants, some collected by the T. D. A. Cockerell expeditions.

PU
Princeton University
Princeton, New Jersey
Most of the large collections of Florissant plants and insects made by the Princeton Expedition of 1877 were housed at Princeton University until 1985, at which time they were transferred to USNM and YPM. About 180 of these specimens are types.

SDNHM
San Diego Natural History Museum
San Diego, California
This is a small collection including one plant holotype.

SDSM
South Dakota School of Mines and Technology
Rapid City, South Dakota
This is a small collection of about 50 plants that are on exhibit.

UCM
University of Colorado Museum
Boulder, Colorado
This is one of the largest collections of Florissant fossils, consisting of about 3,000 specimens of plants, 2,500 insects, and 50 vertebrates. There are 405 published specimens including 349 types. Most of the collection came from Cockerell's Florissant expeditions of 1906–1908.

UCMP
University of California Museum of Paleontology
Berkeley, California
This collection consists of more than 1000 specimens of fossil plants made by Harry D. MacGinitie during his excavations in 1936 and 1937. There are 259 published specimens including 26 types.

UF
Florida Museum of Natural History, University of Florida
Gainesville, Florida
This collection consists of about 200 insects and 300 plants, including 17 published plant specimens. In addition, the museum stores three type specimens of butterflies from the collection of Florissant Fossil Beds National Monument.

UMMP
University of Michigan Museum of Paleontology
Ann Arbor, Michigan
This is a small collection with about 175 specimens, mostly of plants but also with several insects.

USGS
United States Geological Survey
Denver, Colorado
This collection consists of the microscopic slides of pollen and spores studied by Wingate and Nichols.

USNM
National Museum of Natural History (formerly United States National Museum), Smithsonian Institution
Washington, D.C.
This is the largest and best-diversified collection from Florissant, consisting of an estimated 10,400 specimens of plants, insects, fish, birds, and one mammal. It contains 1,045 published specimens including 750 types. The collection came from many different sources and includes specimens collected by the Princeton Expedition of 1877, the Cockerell expeditions, H. F. Wickham, members of the Hayden Survey, Charlotte Hill (who owned part of the fossil beds in the 1870s), and R. D. Lacoe, along with many others. Among the more significant holdings are the plants that were first described by Lesquereux; many of the plants that were cited by Kirchner, Knowlton, or MacGinitie; some of the insects cited by Scudder; most of the beetles that were studied by Wickham; many of the ants described by Carpenter; some of the plants and insects cited by Cockerell; and the fish that were first described by Cope.

UWBM
Burke Museum of Natural History and Culture, University of Washington
Seattle, Washington
This is a small collection of fewer than 100 plants. The pollen and spore samples examined by Leopold and Clay-Poole are intended to be added to this collection.

YPM
Peabody Museum of Natural History, Yale University
New Haven, Connecticut
This is a large collection consisting of approximately 1,600 specimens of plants, insects, fish, and two birds. There are 121 published specimens of which 91 are types. Many of the specimens were collected by the Cockerell expeditions of 1906 and 1907 and the Princeton Expedition of 1877. In addition to part of the collection that was formerly at Princeton University, Yale also now houses the collection that was transferred from the New York Botanical Garden.

WC
Waynesburg College
Waynesburg, Pennsylvania
This is a large collection of more than 3,000 specimens, mostly of plants but also including some insects. The

collection is largely unstudied, and only about five of the specimens have been referenced in publication. Nevertheless, it includes many high-quality, well-preserved specimens, several of which are on exhibit at the college. The collection was made under the direction of Paul R. Stewart, former President of Waynesburg College, between about 1940 and 1971. About 200 specimens from this collection have been on loan to the National Park Service since 1975 for exhibit at Florissant Fossil Beds National Monument, and this exhibit is the largest collection of Florissant fossils on public display.

GENERAL REFERENCES

The following references provide general information about the habitats and classification of the living relatives of the Florissant organisms, and other related topics.

Andrews, H. N. 1980. The fossil hunters: in search of ancient plants. Cornell University Press, Ithaca, N.Y. 420 pages.

Banister, K., and A. Campbell, editors. 1985. The encyclopedia of aquatic life. Facts on File. New York. 349 pages.

Borror, D. J., C. A. Triplehorn, and N. F. Johnson. 1989. An introduction to the study of insects. 6th edition. Saunders College Publishing, Philadelphia. 875 pages.

Cameron, A., illustrator, and C. J. O. Harrison, editor. 1978. Bird families of the world. Harry N. Abrams, New York. 264 pages.

Elias, T. S. 1980. The complete trees of North America: field guide and natural history. Outdoor Life/Nature Books, New York. 948 pages.

Felger, R. S., M. B. Johnson, and M. F. Wilson. 2001. The trees of Sonora, Mexico. Oxford University Press, New York. 391 pages.

Graham, A. 1999. Late Cretaceous and Cenozoic history of North American vegetation. Oxford University Press, New York. 350 pages.

Haq, B. U., and F. W. B. Van Eysinga. 1998. Geologic time table. 5th edition (chart). Elsevier Science, Amsterdam.

Hora, B., editor. 1980. The Oxford encyclopedia of trees of the world. Crescent Books, New York. 288 pages.

Krüssmann, G. 1985. Manual of cultivated conifers. Timber Press, Portland, Ore. 361 pages.

Mabberley, D. J. 1997. The plant-book: a portable dictionary of the vascular plants. 2nd edition. Cambridge University Press, Cambridge. 858 pages.

McGavin, G. C. 2000. Insects, spiders, and other terrestrial arthropods. Dorling Kindersley, New York. 255 pages.

Migdalski, E. C., and G. S. Fichter. 1983. The fresh and salt water fishes of the world. Greenwich House, New York. 316 pages.

O'Toole, C., editor. 1986. The encyclopedia of insects. Facts on File, New York. 143 pages.

Perrins, C. M., and A. L. A. Middleton, editors. 1985. The encyclopedia of birds. Facts on File, New York. 447 pages.

Prothero, D. R. 1994. The Eocene-Oligocene transition: paradise lost. Columbia University Press, New York. 291 pages.

Takhtajan, A. 1997. Diversity and classification of flowering plants. Columbia University Press, New York. 643 pages.

Tidwell, W. D. 1998. Common fossil plants of western North America. Smithsonian Institution Press, Washington, D.C., and London. 299 pages.

BIBLIOGRAPHY

This bibliography is a comprehensive list of the publications that relate to Florissant. These range from publications that are monographic treatments exclusive to Florissant, to those that treat particular taxonomic groups and include at least one reference to a Florissant taxon, to those that deal with aspects of paleoecology, geology, or history. It also includes references that relate to other similar fossil sites, or to general methodologies that have been applied to understanding the Florissant fossils. The earliest scientific publication about Florissant appeared in 1873, and by 1920, most of Florissant's fossil organisms had been named. Many of these early names were later revised, however, and many of the publications of the twentieth century document these changes.

Adams, P. A. 1967. A review of the Mesochrysinae and Nothochrysinae (Neuroptera: Chrysopidae): Bulletin of the Museum of Comparative Zoology at Harvard University 135 (4): 215–238.

Allen, J. A. 1878a. Description of a fossil Passerine bird from the insect-bearing shales of Colorado. Bulletin of the United States Geological and Geographical Survey of the Territories 4:443–445, 1 plate.

Allen, J. A. 1878b. A fossil sparrow-like bird. Nature 18:204–205.

Andrews, H. N. 1936. A new *Sequoioxylon* from Florissant, Colorado. Annals of the Missouri Botanical Garden 23 (3): 439–446, plates 20, 21.

Arnold, C. A. 1936. Some fossil species of *Mahonia* from the Tertiary of eastern and southeastern Oregon. Contributions from the Museum of Paleontology, University of Michigan 5 (4): 57–66, 3 plates.

Askevold, I. S. 1990. Classification of Tertiary fossil Donaciinae of North America and their implications about the evolution of Donaciinae (Coleoptera: Chrysomelidae). Canadian Journal of Zoology 68 (10): 2135–2145.

Axelrod, D. I. 1986. Cenozoic history of some western American pines. Annals of the Missouri Botanical Garden 73:565–641.

Axelrod, D. I. 1987. The late Oligocene Creede flora, Colorado. University of California Publications, Geological Sciences 130:i–x, 1–235.

Axelrod, D. I. 1997. Paleoelevation estimated from Tertiary floras. International Geology Review 39:1124–1133.

Bather, F. A. 1909. Visit to the Florissant exhibition in the British Museum (Natural History). Proceedings of the Geologists' Association 21, pt. 3: 159–165.

Becker, H. F. 1961. Oligocene plants, upper Ruby River Basin, Montana. Geological Society of America Memoir 82:1–127, 32 plates.

Becker, H. F. 1969. Fossil plants of the Tertiary Beaverhead Basins in southwestern Montana. Palaeontographica, Abt. B 127:1–142, 44 plates.

Becker-Migdisova, E. E. 1985. Fossil insects Psyllomorpha. Trudy Paleontologicheskogo Instituta, Akademiia Nauk SSSR 206:83–86. [In Russian.]

Beetle, A. A. 1958. *Piptochaetium* and *Phalaris* in the fossil record. Bulletin of the Torrey Botanical Club 85 (3): 179–181.

Benson, R. B. 1942. Blasticotomidae in the Miocene of Florissant, Colorado (Hymenoptera: Symphyta). Psyche 49 (3-4): 47–48.

Bequaert, J. C. 1930. Tsetse flies—past and present (Diptera: Muscoidea). Entomological News 41:158–164, 202–203, 227–233.

Bequaert, J. C. 1932. The Nemestrinidae (Diptera) in the V. v. Röder collection. Zoologischer Anzeiger, Band 100:13–33.

Bequaert, J. C. 1947. Catalogue of recent and fossil Nemestrinidae of North America north of Mexico. Psyche 54:194–207.

Bequaert, J. C., and F. M. Carpenter. 1936. The Nemestrinidae of the Miocene of Florissant, Colorado and their relations to the recent fauna. Journal of Paleontology 10 (5): 395–409.

Britton, E. G., and A. Hollick. 1907. American fossil mosses, with description of a new species from Florissant, Colorado. Bulletin of the Torrey Botanical Club 34 (3): 139–149, plate 9.

Britton, E. G., and A. Hollick. 1915. A new American fossil moss. Bulletin of the Torrey Botanical Club 42 (1): 9–10.

Brown, F. M. 1976. *Oligodonta florissantensis,* gen. nov., sp. nov. (Lepidoptera: Pieridae). Bulletin of the Allyn Museum 37:1–4.

Brown, F. M. 1981. A note about Florissant fossil insects. Entomological News 92 (4): 165–166.

Brown, F. M. 1985. Four undescribed Oligocene craneflies from Florissant, Colorado (Diptera: Tipulidae). Insecta Mundi 1 (2): 98–100.

Brown, F. M. 1986a. *Lithoserix williamsi* (Siricidae: Hymenoptera) a newly recognized fossil horntail from Florissant, Colorado. Insecta Mundi 1 (3): 119–120.

Brown, F. M. 1986b. The tsetse flies of Florissant, Colorado. Bulletin of Pikes Peak Research Station 1:1–14, 2 plates.

Brown, R. W. 1937. Additions to some fossil floras of the Western United States. U.S. Geological Survey Professional Paper 186-J:163–206, plates 45–63.

Brues, C. T. 1906. Fossil parasitic and phytophagous Hymenoptera from Florissant, Colorado. Bulletin of the American Museum of Natural History 22:491–498 (art. 29).

Brues, C. T. 1908a. Two fossil Phoridae from the Miocene shales of Florissant, Colorado. Bulletin of the American Museum of Natural History 24:273–275 (art. 17).

Brues, C. T. 1908b. New phytophagous Hymenoptera from the Tertiary of Florissant, Colorado. Bulletin of the Museum of Comparative Zoology at Harvard College 51 (10): 259–276.

Brues, C. T. 1910a. The parasitic Hymenoptera of the Tertiary of Florissant, Colorado. Bulletin of the Museum of Comparative Zoology at Harvard College 54 (1): 1–125, plate 1.

Brues, C. T. 1910b. Some notes of the geological history of the parasitic Hymenoptera. Journal of the New York Entomological Society 18 (1): 1–22.

Brues, C. T., and B. B. Brues. 1908. A new fossil grass from the Miocene of Florissant, Colorado (*Melica primaeva*). Bulletin of the Wisconsin Natural History Society 6 (3-4): 170–171.

Buckton, G. B. 1883. Monograph of the British Aphides, Vol. 4:i–ix, 1–228. Ray Society, London.

Calvert, P. P. 1913. The fossil odonate *Phenacolestes,* with a discussion of the venation of the legion Podagrion Selys. Proceedings of the Academy of Natural Sciences of Philadelphia 65, pt. 2:225–272, plate 14.

Carpenter, F. M. 1930. The fossil ants of North America. Bulletin of the Museum of Comparative Zoology at Harvard College 70 (1): 1–66, plates 1–11.

Carpenter, F. M. 1931. The affinities of *Holcorpa maculosa* Scudder and other Tertiary Mecoptera, with descriptions of new genera. Journal of the

New York Entomological Society 39:405–414.

Carpenter, F. M. 1935a. A new name for *Lithomyrex* Carp. (Hymenoptera). Psyche 42:91.

Carpenter, F. M. 1935b. Tertiary insects of the Family Chrysopidae. Journal of Paleontology 9 (3): 259–271.

Carpenter, F. M. 1936. Revision of the Nearctic Raphidiodea (recent and fossil). Proceedings of the American Academy of Arts and Sciences 71 (2): 89–157, plates 1–2.

Carpenter, F. M. 1943. Osmylidae of the Florissant shales, Colorado (Insecta-Neuroptera). American Journal of Science 241:753–760, 1 plate.

Carpenter, F. M. 1959. Fossil Nemopteridae (Neuroptera). Psyche 66 (1-2): 20–24, 1 plate.

Carpenter, F. M. 1985. Substitute names for some extinct genera of fossil insects. Psyche 92 (4): 575–583.

Carpenter, F. M. 1992. Superclass Hexapoda. *In* Arthropoda 4, vols. 3 and 4, pt. R of Treatise on invertebrate paleontology, edited by R. L. Kaesler. Geological Society of America, Boulder, Colorado, and University of Kansas, Lawrence. 655 pages.

Carpenter, F. M., and F. M. Hull. 1939. The fossil Pipunculidae. Bernstein-Forschungen, Heft 4:8–17.

Carpenter, G. H. D. 1919. Protozoal parasites in Cainozoic times. Nature 103 (no. 2577): 46.

Chandler, R. M. 1999. Fossil birds of Florissant, Colorado: with a description of a new species of cuckoo. *In* National Park Service Paleontological Research. Volume 4, edited by V. L. Santucci. Geologic Resources Division Technical Report NPS/NRGRD/GRDTR-99/03:49–53.

Chaney, R. W. 1951. A revision of fossil *Sequoia* and *Taxodium* in western North America based on the recent discovery of *Metasequoia*. Transactions of the American Philosophical Society 40, pt. 3:171–263.

Cockerell, T. D. A. 1906a. The bees of Florissant, Colorado. Bulletin of the American Museum of Natural History 22:419–455 (art. 25).

Cockerell, T. D. A. 1906b. A fossil cicada from Florissant, Colorado. Bulletin of the American Museum of Natural History 22:457–458 (art. 26).

Cockerell, T. D. A. 1906c. The fossil Mollusca of Florissant, Colorado. Bulletin of the American Museum of Natural History 22:459–462 (art. 27).

Cockerell, T. D. A. 1906d. Fossil saw-flies from Florissant, Colorado. Bulletin of the American Museum of Natural History 22:499–501 (art. 30).

Cockerell, T. D. A. 1906e. A fossil water-bug (*Corixa florisantella* n. sp.). Canadian Entomologist 38:209.

Cockerell, T. D. A. 1906f. The fossil fauna and flora of the Florissant (Colorado) shales. The University of Colorado Studies 3:157–176.

Cockerell, T. D. A. 1906g. Fossil Hymenoptera from Florissant, Colorado. Bulletin of the Museum of Comparative Zoology at Harvard College 50 (2):33–58.

Cockerell, T. D. A. 1906h. A new Tertiary *Planorbis*. The Nautilus 19:100–101.

Cockerell, T. D. A. 1906i. Fossil plants from Florissant, Colorado. Bulletin of the Torrey Botanical Club 33:307–312.

Cockerell, T. D. A. 1906j. *Rhus* and its allies. Torreya 6 (1): 11–12

Cockerell, T. D. A. 1907a. An enumeration of the localities in the Florissant basin, from which fossils were obtained in 1906. Bulletin of the American Museum of Natural History 23:127–132 (art. 4).

Cockerell, T. D. A. 1907b. Fossil dragonflies from Florissant, Colorado. Bulletin of the American Museum of Natural History 23:133–139 (art. 5).

Cockerell, T. D. A. 1907c. Some fossil arthropods from Florissant, Colorado. Bulletin of the American Museum of Natural History 23:605–616 (art. 24).

Cockerell, T. D. A. 1907d. Some Coleoptera and Arachnida from Florissant, Colorado. Bulletin of the American Museum of Natural History 23:617–621 (art. 25).

Cockerell, T. D. A. 1907e. A fossil caterpillar. The Canadian Entomologist 39:187–188.

Cockerell, T. D. A. 1907f. A fossil butterfly of the genus *Chlorippe*. The Canadian Entomologist 39 (2): 361–363.

Cockerell, T. D. A. 1907g. A fossil tortricid moth. The Canadian Entomologist 39 (12): 416.

Cockerell, T. D. A. 1907h. A fossil tsetse fly in Colorado. Nature 76:414.

Cockerell, T. D. A. 1907i. A Miocene wasp. Nature 77 (no. 1987): 80.

Cockerell, T. D. A. 1907j. A new Zonitoid shell from

the Miocene, Florissant, Colorado. The Nautilus 21 (8): 89.

Cockerell, T. D. A. 1907k. Some old-world types of insects in the Miocene of Colorado. Science 26:446–447.

Cockerell, T. D. A. 1907l. A redwood described as a moss. Torreya 7 (10): 203–204.

Cockerell, T. D. A. 1907m. A fossil honey-bee. Entomologist 40:227–229.

Cockerell, T. D. A. 1908a. Descriptions of Tertiary insects (I). American Journal of Science, ser. 4, 25:51–52 (art. 4).

Cockerell, T. D. A. 1908b. Descriptions of Tertiary insects (II). American Journal of Science, ser. 4, 25:227–232 (art. 25).

Cockerell, T. D. A. 1908c. Descriptions of Tertiary insects (III). American Journal of Science, ser. 4, 25:309–312 (art. 33).

Cockerell, T. D. A. 1908d. Descriptions of Tertiary Plants (I). American Journal of Science, ser. 4, 26:65–68 (art. 5).

Cockerell, T. D. A. 1908e. Descriptions of Tertiary insects (IV). American Journal of Science, ser. 4, 26:69–75 (art. 6).

Cockerell, T. D. A. 1908f. Descriptions of Tertiary plants (II). American Journal of Science, ser. 4, 26:537–544 (art. 50).

Cockerell, T. D. A. 1908g. Fossil insects from Florissant, Colorado. Bulletin of the American Museum of Natural History 24:59–69 (art. 3).

Cockerell, T. D. A. 1908h. The fossil flora of Florissant, Colorado. Bulletin American Museum of Natural History 24:71–110 (art. 4).

Cockerell, T. D. A. 1908i. Some results of the Florissant expedition of 1908. The American Naturalist 42 (no. 501): 569–581.

Cockerell, T. D. A. 1908j. Descriptions and records of bees (19). Annals and Magazine of Natural History, British Museum of Natural History, London, ser. 8, 1:337–344 (art. 53).

Cockerell, T. D. A. 1908k. List of fossil Histeridae from the Tertiary strata [*In* G. Lewis, On new species of Histeridae and notices of others]. Annals and Magazine of Natural History, British Museum of Natural History, London,, ser. 8, 2:160–162.

Cockerell, T. D. A. 1908l. Descriptions and records of bees (20). Annals and Magazine of Natural History, ser. 8, 2:323–334.

Cockerell, T. D. A. 1908m. A fossil leaf-cutting bee. The Canadian Entomologist 40 (1): 31–32.

Cockerell, T. D. A. 1908n. Fossil Chrysopidae. The Canadian Entomologist 40 (3): 90–91.

Cockerell, T. D. A. 1908o. Two fossil Diptera. The Canadian Entomologist 40 (6): 173–175, plate 4.

Cockerell, T. D. A. 1908p. Fossil Osmylidae (Neuroptera) in America. The Canadian Entomologist 40 (10): 341–342.

Cockerell, T. D. A. 1908q. The first American fossil mantis (*Lithophotina floccose*). The Canadian Entomologist 40 (10): 343–344.

Cockerell, T. D. A. 1908r. A fossil orthopterous insect with the media and cubitus fusing. Entomological News 19 (3): 126–128.

Cockerell, T. D. A. 1908s. A dragon-fly puzzle and its solution. Entomological News 19:455–459.

Cockerell, T. D. A. 1908t. The Dipterous Family Nemestrinidae. Transactions of the American Entomological Society 34:247–253, plate 16.

Cockerell, T. D. A. 1908u. Another fossil nemestrinid fly (*Hironeura occuitar*), Florissant, Colorado. Transactions of the Entomological Society of America 34:254.

Cockerell, T. D. A. 1908v. A fossil fly of the family Blepharoceridae (*Philorites johannseni* n. g., n. sp.). The Entomologist 41:262–265.

Cockerell, T. D. A. 1908w. Fossil Aphididae from Florissant, Colorado. Nature 78 (no. 2023): 318–319.

Cockerell, T. D. A. 1908x The Miocene species of Lymnaea. The Nautilus 22 (7): 69–70.

Cockerell, T. D. A. 1908y. Florissant: A Miocene Pompeii. The Popular Science Monthly 74 (2): 112–126.

Cockerell, T. D. A. 1908z. The fossil sawfly *Perga coloradensis*. Science 27 (681): 113–114.

Cockerell, T. D. A. 1908zz. Fossil Cercopidae (Homoptera). Bulletin of the Wisconsin Natural History Society 6 (1-2): 35–38.

Cockerell, T. D. A. 1909a. Descriptions of Tertiary insects (V): some new Diptera. American Journal of Science, ser. 4, 27:53–58 (art. 3).

Cockerell, T. D. A. 1909b. Descriptions of Tertiary insects (VI). American Journal of Science, ser. 4, 27:381–387 (art. 32).

Cockerell, T. D. A. 1909c. Descriptions of Tertiary insects (VII). American Journal of Science, ser. 4,

28:283–286 (art. 30).

Cockerell, T. D. A. 1909d. Fossil Diptera from Florissant, Colorado. Bulletin of the American Museum of Natural History 26:9–12 (art. 2).

Cockerell, T. D. A. 1909e. Fossil insects from Florissant, Colorado. Bulletin of the American Museum of Natural History 26:67–76, plate 16 (art. 7).

Cockerell, T. D. A. 1909f. A catalogue of the generic names based on American insects and arachnids from the Tertiary rocks, with indications of the type species. Bulletin of the American Museum of Natural History 26:77–86 (art. 8).

Cockerell, T. D. A. 1909g. Two fossil Chrysopidae. The Canadian Entomologist 41 (7): 218–219.

Cockerell, T. D. A. 1909h. New North American bees. The Canadian Entomologist 41 (11): 393–395.

Cockerell, T. D. A. 1909i. Two fossil bees. Entomological News 20 (4): 159–161.

Cockerell, T. D. A. 1909j. New fossil insects from Florissant, Colorado. Annals of the Entomological Society of America 2:251–256, plate 28.

Cockerell, T. D. A. 1909k. Fossil Insects from Colorado. The Entomologist 42 (554): 170–174.

Cockerell, T. D. A. 1909l. Fossil Euphorbiaceae, with a note on Saururaceae. Torreya 9 (6): 117–119.

Cockerell, T. D. A. 1909m. Two new fossil plants from Florissant, Colorado. Torreya 9 (9): 184–185.

Cockerell, T. D. A. 1910a. Descriptions of Tertiary plants (III): a *Sorbus* from Florissant, considered to be a hybrid. American Journal of Science 29:76–78 (art. 7).

Cockerell, T. D. A. 1910b. Fossil insects and a crustacean from Florissant, Colorado. Bulletin of the American Museum of Natural History 28:275–288 (art. 25).

Cockerell, T. D. A. 1910c. The Miocene trees of the Rocky Mountains. The American Naturalist 44:31–47.

Cockerell, T. D. A. 1910d. The fossil Crabronidae. The Entomologist 43:60–61.

Cockerell, T. D. A. 1910e. A Tertiary leaf-cutting bee. Nature 82 (no. 2102): 429.

Cockerell, T. D. A. 1910f. *Magnolia* at Florissant. Torreya 10 (3): 64–65.

Cockerell, T. D. A. 1910g. Notes on the genus *Sambucus*. Torreya 10 (6): 125–128.

Cockerell, T. D. A. 1910h. A fossil fig. Torreya 10 (10): 222–224.

Cockerell, T. D. A. 1910i. The Miocene flora. In The age of mammals in Europe, Asia, and North America, edited by H. F. Osborn, 282–285. Macmillan, New York.

Cockerell, T. D. A. 1911a. Fossil insects from Florissant, Colorado. Bulletin of the American Museum of Natural History 30:71–82, plate 3 (art. 6).

Cockerell, T. D. A. 1911b. Descriptions and records of bees (39). Annals and Magazine of Natural History, ser. 8, 7:225–237.

Cockerell, T. D. A. 1911c. Note on *Lymnaea florissantica*. The Nautilus 25:24.

Cockerell, T. D. A. 1911d. Scudder's work on fossil insects. Psyche 18 (6): 181–186.

Cockerell, T. D. A. 1911e. Fossil flowers and fruits. Torreya 11 (11): 234–236.

Cockerell, T. D. A. 1912. A fossil *Raphidia* (Neur., Planip.). Entomological News 23:215–216.

Cockerell, T. D. A. 1913a. The fauna of the Florissant, Colorado shales. American Journal of Science, ser. 4, 36:498–500 (art. 44).

Cockerell, T. D. A. 1913b. Some fossil insects from Florissant, Colorado. The Canadian Entomologist 45:229–233.

Cockerell, T. D. A. 1913c. The first fossil anthomyid fly from Florissant (Dipt.). Entomological News 24:295–296.

Cockerell, T. D. A. 1913d. Remarks on fossil insects. Proceedings of the Entomological Society of Washington 15:123–126.

Cockerell, T. D. A. 1913e. The first fossil mydaid fly. Entomologist 46:207–208.

Cockerell, T. D. A. 1913f. A fossil asilid fly from Colorado. Entomologist 46:213–214.

Cockerell, T. D. A. 1913g. The genus *Phryganea* (Trichoptera) in the Florissant shales. Psyche 20:95–96.

Cockerell, T. D. A. 1913h. Fossil flowers and fruits (III). Torreya 13 (4): 75–77.

Cockerell, T. D. A. 1913i. Some fossil insects from Florissant, Colorado. Proceedings of the United States National Museum 44 (no. 1955): 341–346, plate 56.

Cockerell, T. D. A. 1913j. Two fossil insects from Florissant, Colorado, with a discussion of the venation of the Aeshnine dragon-flies.

Proceedings of the United States National Museum 45 (no. 2000): 577–583.

Cockerell, T. D. A. 1914a. The fossil and recent Bombyliidae compared. Bulletin of the American Museum of Natural History 33:229–236 (art. 18).

Cockerell, T. D. A. 1914b. Three Diptera from the Miocene of Colorado. The Canadian Entomologist 46:101–102.

Cockerell, T. D. A. 1914c. The Florissant fossil insects. *In* Proceedings of the Colorado Scientific Society, edited by J. Henderson. Recent Progress in Colorado Paleontology and Stratigraphy 11:8–10.

Cockerell, T. D. A. 1914d. The fossil Orthoptera of Florissant, Colorado. The Entomologist 47:32–34.

Cockerell, T. D. A. 1914e. New and little-known insects from the Miocene of Florissant, Colorado. The Journal of Geology 22:714–724.

Cockerell, T. D. A. 1914f. Miocene fossil insects. Proceedings of the Academy of Natural Sciences of Philadelphia 66: 634–648.

Cockerell, T. D. A. 1914g. Two new plants from the Tertiary rocks of the west. Torreya 14 (8): 135–137.

Cockerell, T. D. A. 1915a. Briefer articles: Notes on orchids. Botanical Gazette 59:331–333.

Cockerell, T. D. A. 1915b. A fossil fungus-gnat (*Mycetophila bradenae*). The Canadian Entomologist 47:159.

Cockerell, T. D. A. 1915c. *Equisetum* in the Florissant Miocene. Torreya 15 (12): 265–267.

Cockerell, T. D. A. 1916a. Colorado a million years ago. American Museum Journal 16 (7): 442–450.

Cockerell, T. D. A. 1916b. Two Diptera of the genus *Rhamphomyia* from Colorado. The Canadian Entomologist 48:123–124.

Cockerell, T. D. A. 1916c. The third fossil tsetse-fly. Nature 98 (no. 2448): 70.

Cockerell, T. D. A. 1916d. Some American fossil insects. Proceedings of the United States National Museum 51 (no. 2146): 89–106, plate 2.

Cockerell, T. D. A. 1917a. A fossil tsetse fly and other Diptera from Florissant, Colorado. Proceedings of the Biological Society of Washington 30:19–21.

Cockerell, T. D. A. 1917b. Descriptions of fossil insects. Proceedings of the Biological Society of Washington 30:79–82.

Cockerell, T. D. A. 1917c. Fossil insects. Annals of the Entomological Society of America 10 (1): 1–22.

Cockerell, T. D. A. 1917d. New Tertiary insects. Proceedings of the United States National Museum 52:373–384, plate 31.

Cockerell, T. D. A. 1917e. Some fossil insects from Florissant, Colorado. Proceedings of the United States National Museum 53 (no. 2210): 389–392.

Cockerell, T. D. A. 1919a. Some fossil parasitic Hymenoptera. American Journal of Science, ser. 4, 47:376–380 (art. 25).

Cockerell, T. D. A. 1919b. The antiquity of the Ichneumonidae. The Entomologist 52 (no. 673): 121–122.

Cockerell, T. D. A. 1919c. *Glossina* and the extinction of Tertiary mammals. Nature 103 (no. 2588): 265.

Cockerell, T. D. A. 1922a. A fossil moth from Florissant, Colorado. American Museum Novitates (34): 1–2.

Cockerell, T. D. A. 1922b The fossil sawflies of Florissant, Colorado. The Entomologist 55 (706): 49–50.

Cockerell, T. D. A. 1922c. A fossil buttercup (*Ranunculus florssantensis*). Nature 109 (no. 2724): 42–43.

Cockerell, T. D. A. 1922d A new genus of fossil Liliaceae. Bulletin of the Torrey Botanical Club 49 (7): 211–213.

Cockerell, T. D. A. 1923a. Two fossil Hymenoptera from Florissant (Vespidae, Megachilidae). Entomological News 34:270–271.

Cockerell, T. D. A. 1923b. A new genus of mayflies from the Miocene of Florissant, Colorado. Psyche 30 (5): 170–172.

Cockerell, T. D. A. 1924a. Fossil insects. Entomological News 35:28–30.

Cockerell, T. D. A. 1924b. Fossil ichneumons believed to have been parasitic on sawflies. The Entomologist 57:9–10.

Cockerell, T. D. A. 1924c. Fossil insects in the United States National Museum. Proceedings of the United States National Museum 64:1–15, plates 1, 2 (no. 2503, art. 13).

Cockerell, T. D. A. 1925. A new fossil dragonfly from Florissant. The Entomologist 58:205–206.

Cockerell, T. D. A. 1926a. Some Tertiary fossil insects. Annals and Magazine of Natural History, ser. 9, 18:313–324 (no. 106, art. 40).

Cockerell, T. D. A. 1926b. A fossil Orthopterous insect formerly referred to Mecoptera. Proceedings of the Entomological Society of Washington 28 (6): 142.

Cockerell, T. D. A. 1926c. A new fossil moth from

Florissant. Psyche 33 (1): 16–17.

Cockerell, T. D. A. 1926d. Hunting fossil insects. Scientific American 134:264–265.

Cockerell, T. D. A. 1926e. A Miocene *Orontium* (Araceae). Torreya 26 (4): 69.

Cockerell, T. D. A. 1927a. Hymenoptera and a caddis larva from the Miocene of Colorado. Annals and Magazine of Natural History, ser. 9, 20:429–435.

Cockerell, T. D. A. 1927b. Fossil insects in the British Museum. Annals and Magazine of Natural History, ser. 9, 20:585–594.

Cockerell, T. D. A. 1927c. New name for a trichopterous genus. The Entomologist 60:184–185.

Cockerell, T. D. A. 1927d. A supposed fossil catmint. Torreya 27 (3): 54.

Cockerell, T. D. A. 1930. The Miocene shales of Florissant. Journal of the Colorado-Wyoming Academy of Science 1 (2): 24.

Cockerell, T. D. A. 1937a. Recollections of a naturalist (IV): the amateur botanist. Bios (Beta Beta Beta) 8:12–18.

Cockerell, T. D. A. 1937b. Recollections of a naturalist (V): fossil insects. Bios (Beta Beta Beta) 8 (2): 51–56.

Cockerell, T. D. A. 1941. Some Tertiary insects (Hymenoptera) from Colorado. American Journal of Science 239:354–356.

Cockerell, T. D. A., and C. Custer. 1925. A new fossil *Inocellia* (Neuroptera) from Florissant. The Entomologist 58:295–297.

Cokendolpher, J. C., and J. E. Cokendolpher. 1982. Reexamination of the Tertiary harvestmen from the Florissant Formation, Colorado (Arachnida: Opiliones: Palpatores). Journal of Paleontology 56 (5): 1213–1217.

Cooper, K. W. 1941. *Davispia bearcreekensis* Cooper, a new cicada from the Paleocene, with a brief review of the fossil Cicadidae. American Journal of Science 239:286–304.

Cope, E. D. 1874. Supplementary notices of fishes from the freshwater Tertiaries of the Rocky Mountains. Bulletin of the United States Geological and Geographical Survey of the Territories, I, ser. 1, no. 2:49–51.

Cope, E. D. 1875. On the fishes of the Tertiary shales of the South Park. Bulletin of the United States Geological and Geographical Survey of the Territories, I, ser. 2, no. 1: 3–5.

Cope, E. D. 1878. Descriptions of fishes from the Cretaceous and Tertiary deposits west of the Mississippi river. Bulletin of the United States Geological and Geographical Survey of the Territories 4:67–77 (art. 2).

Cope, E. D. 1882. On a wading bird from the Amyzon Shales. Bulletin of the United States Geological Survey 6:83–85 (art. 3).

Cope, E. D. 1883. The Vertebrata of the Tertiary formations of the West. Report of the United States Geological Survey of the Territories 3:1–1009, 134 plates [see particularly "The Amyzon Shales," Supplement to Part First].

Cross, W. 1894. Pikes Peak Folio, Colorado. Geologic atlas of the United States, Vol. 7. United States Geological Survey, Washington, D.C.

Demoulin, G. 1970. Contribution à la connaissance des Éphéméroptères du Miocène, I, *Siphlurites explanatus* Cockerell. Bulletin de l'Institut royal des sciences naturelles de Belgique 46 (5): 1–4.

Durden, C. J. 1966. Oligocene lake deposits in central Colorado and a new fossil insect locality. Journal of Paleontology 40 (1): 215–219.

Edwards, W. N. 1927. The occurrence of *Koelreuteria* (Sapindaceae) in Tertiary rocks. Annals and Magazine of Natural History, ser. 9, 20:109–112.

Eklund, R., Jr. 1968. The Florissant fossil beds. Earth Science (September-October): 217–218.

Emerson, A. E. 1933. A revision of the genera of fossil and recent Termopsinae (Isoptera). University of California Publications in Entomology 6 (6): 165–196.

Emerson, A. E. 1969. A revision of the Tertiary fossil species of the Kalotermitidae (Isoptera). American Museum Novitates 2359:1–57.

Emmel, T. C., M. C. Minno, and B. A. Drummond. 1992. Fossil butterflies: a guide to the fossil and present-day species of central Colorado. Stanford University Press, Stanford, Calif.

Epis, R. C. 1969. Proposed Florissant Fossil Beds National Monument. Mines Magazine 59 (5): 10–13.

Epis, R. C., and C. E. Chapin. 1974. Stratigraphic nomenclature of the Thirtynine Mile volcanic field, central Colorado. Geological Survey Bulletin 1395-C:C1–C23.

Epis, R. C., and C. E. Chapin. 1975. Geomorphic and

tectonic implications of the post-Laramide, late Eocene erosion surface in the southern Rocky Mountains. *In* Cenozoic history of the southern Rocky Mountains, edited by B. F. Curtis. Geological Society of America Memoir 144:45–74.

Evanoff, E. 1999. Fossil mammals and biting insects from the upper Eocene Florissant Formation of Colorado. Abstracts of papers, 59th annual meeting, Society of Vertebrate Paleontology. Journal of Vertebrate Paleontology 19 (suppl. to no. 3): 43A.

Evanoff, E., and P. M. deToledo. 1993. New fossil mammals found at Florissant Fossil Beds National Monument. Park Science 13 (4): 13.

Evanoff, E., K. M. Gregory-Wodzicki, and K. R. Johnson, eds. 2001. Fossil flora and stratigraphy of the Florissant Formation, Colorado. Proceedings of the Denver Museum of Nature & Science, ser. 4, no. 1.

Evanoff, E., W. C. McIntosh, and P. C. Murphey. 2001. Stratigraphic summary and ^{40}Ar/^{39}Ar geochronology of the Florissant Formation, Colorado. *In* Fossil flora and stratigraphy of the Florissant Formation, Colorado, edited by E. Evanoff, K. M. Gregory-Wodzicki, and K. R. Johnson. Proceedings of the Denver Museum of Nature & Science, ser. 4, 1:1–16.

Evenhuis, N. L. 1984. A reassessment of the taxonomic position of the fossil species *Protophthiria palpalis* and *P. atra* (Diptera: Bombyliidae). International Journal of Entomology 26 (1-2): 157–161.

Forest, C. E., P. Molnar, and K. A. Emanuel. 1995. Paleoaltimetry from energy conservation principles. Nature 374:347–350.

Gamer, E. E. 1965. The fossil beds of Florissant. National Parks Magazine, July 1965, 16–19.

Gazin, C. L. 1935. A marsupial from the Florissant Beds (Tertiary) of Colorado. Journal of Paleontology 9 (1): 57–62.

Grande, L., and W. E. Bemis. 1998. A comprehensive phylogenetic study of amiid fishes (Amiidae) based on comparative skeletal anatomy: an empirical search for interconnected patterns of natural history. Society of Vertebrate Paleontology Memoir 4:1–690.

Gregory, K. M. 1992. Late Eocene paleoaltitude, paleoclimate, and paleogeography of the Front Range region, Colorado. Ph. D. dissertation, University of Arizona, Tucson. 246 pages.

Gregory, K. M. 1994. Paleoclimate and paleoelevation of the 35 Ma Florissant flora, Front Range, Colorado. Palaeoclimates 1:23–57.

Gregory, K. M., and C. G. Chase. 1992. Tectonic significance of paleobotanically estimated climate and altitude of the late Eocene erosion surface, Colorado. Geology 20:581–585.

Gregory, K. M, and C. G. Chase. 1994. Tectonic and climatic significance of a late Eocene low-relief, high-level geomorphic surface, Colorado. Journal of Geophysical Research 99 (B10): 20,141–20,160.

Gregory, K. M., and W. C. McIntosh. 1996. Paleoclimate and paleoelevation of the Oligocene Pitch-Pinnacle flora, Sawatch Range, Colorado. GSA Bulletin 108 (5): 545–561.

Gregory-Wodzicki, K. M. 2001. Paleoclimatic implications of tree-ring growth characteristics of 34. 1 Ma *Sequoioxylon pearsallii* from Florissant, Colorado. *In* Fossil flora and stratigraphy of the Florissant Formation, Colorado, edited by E. Evanoff, K. M. Gregory-Wodzicki, and K. R. Johnson. Proceedings of the Denver Museum of Nature & Science, ser. 4, 1:163–186.

Grimaldi, D. A. 1992. Vicariance biogeography, geographic extinctions, and the North American Oligocene tsetse flies. *In* Extinction and phylogeny, edited by M. Novacek and Q. Wheeler, 178–204. Columbia University Press, New York.

Harding, I. C., and L. S. Chant. 2000. Self-sedimented diatom mats as agents of exceptional fossil preservation in the Oligocene Florissant lake beds, Colorado, United States. Geology 28 (3): 195–198.

Hartman, J. H. 1998. The stratigraphy of Mesozoic and early Cenozoic nonmarine mollusks of Colorado. Proceedings of the Denver Museum of Natural History, ser. 3, 14:1–15.

Hascall, A. P. 1988. Stratigraphic palynology, vegetation dynamics and paleoecology of the Florissant lakebeds (Oligocene), Colorado. Master's thesis, Michigan State University, East Lansing. 149 pages.

Hatch, M. H. 1927. Studies on the Silphinæ. Journal of the New York Entomological Society 35:331–370, plate 16.

Heie, O. E. 1967. Studies on fossil aphids (Homoptera: Aphidoidea). Spolia Zoologica Musei Hauniensis 26:1–273.

Heilprin, A. 1896. The stone forest of Florissant. Appelton's Popular Science Monthly 49:479–484.

Henderson, J. 1906. The Tertiary lake basin of Florissant, Colorado. University of Colorado Studies 3 (3): 145–156.

Henry, T. W., E. Evanoff, D. Grenard, H. W. Meyer, and J. A. Pontius. 1996. Geology of the Gold Belt Back Country Byway, south-central Colorado. *In* Geologic excursions to the Rocky Mountains and beyond, edited by R. A. Thompson, M. R. Hudson, and C. L. Pillmore. Field trip guidebook for the 1996 annual meeting of the Geological Society of America, Denver, Colorado, October 28–31. Colorado Geological Survey Special Publication 44. 48 pages on CD-ROM.

Hequist, K. J. 1961. Notes on Cleonymidae (Hymenoptera: Chalcidoidea), I. Entomologisk Tidskrift 82:91–110.

Hollick, A. 1894. Fossil salvinias, including description of a new species. Torrey Botanical Club Bulletin 21:253–257, plate 205.

Hollick, A. 1907. Description of a new Tertiary fossil flower from Florissant, Colorado. Torreya 7:182–184.

Hollick, A. 1909. A new genus of fossil Fagaceae from Colorado. Torreya 9 (1): 1–3.

Howe, M. A., and A. Hollick. 1922. A new American fossil hepatic. Bulletin of the Torrey Botanical Club 49 (7): 207–209.

Hull, F. M. 1945. A revisional study of the fossil Syrphidae. Bulletin of the Museum of Comparative Zoology at Harvard College 95 (3): 251–355, plates 1–13.

Hull, F. M. 1949. The morphology and inter-relationship of the genera of syrphid flies, recent and fossil. Transactions of the Zoological Society of London 26, pt. 4:257–408.

Hull, F. M. 1957. Tertiary flies from Colorado and the Baltic amber. Psyche 64 (2): 37–45, plates 2–4.

Hull, F. M. 1960. A new genus and four new species of fossil Diptera from Montana and Colorado. Contributions from the Museum of Paleontology, University of Michigan 15 (11): 269–279, plates 1–4.

Hungerford, H. B. 1932. Concerning a fossil water bug from the Florissant (Nepidae). University of Kansas Science Bulletin 20 (19): 327–331.

Hutchinson, R. M., and K. E. Kolm. 1987. The Florissant Fossil Beds National Monument, Teller County, Colorado. Geological Society of America Centennial Field Guide 2:329–330.

Jacobson, S. R., and D. J. Nichols. 1981. Florissant: an ancient flora preserved. Garden, March-April 1981, 19–23.

James, M. T. 1937. A preliminary review of certain families of Diptera from the Florissant Miocene beds. Journal of Paleontology 11 (3): 241–247.

James, M. T. 1939. A preliminary review of certain families of Diptera from the Florissant Miocene beds II. Journal of Paleontology 13 (1): 42–48.

Johannsen, O. A. 1912. A Tertiary fungus gnat. American Journal of Science, ser. 4, 34:140.

Kasparyan D. R., and A. P., Rasnitsyn. 1992. The systematic position of the Oligocene *Lithoserix williamsi* Brown, (Hymenoptera: Ichneumonidae) from Colorado (USA). Paleontological Journal 26 (3): 134–135.

Kennedy, C. H. 1925. New genera of Megapodagrioninae, with notes on the subfamily. Bulletin of the Museum of Comparative Zoology at Harvard College 67:291–312, plate 1.

Kevan, D. K. McE., and D. C. Wighton. 1983. Further observations on North American Tertiary orthopteroids (Insecta: Grylloptera). Canadian Journal of Earth Science 20:217–224.

Kingsley, J. S. 1911. Samuel Hubbard Scudder. Psyche 18 (6): 175–178.

Kinsey, A. C. 1919. Fossil Cynipidae. Psyche 26 (2): 44–49.

Kirchner, W. C. G. 1898 Contribution to the fossil flora of Florissant, Colorado. Transactions of the Academy of Science of St. Louis 8:161–188, plates 11–15.

Kirkaldy, G. W. 1910. Three new Hemiptera-Heteroptera from the Miocene of Colorado. Entomological News 21:129–131.

Knowlton, F. H. 1916. A review of the fossil plants in the United States National Museum from the Florissant Lake Beds at Florissant, Colorado, with a description of new species and a list of type-specimens. Proceedings of the United States National Museum 51 (no. 2151): 241–297, plates 12–27.

Kohl, M. F., and J. S. McIntosh, eds. 1997. Discovering dinosaurs in the old West: the field journals of Arthur Lakes. Smithsonian Institution Press, Washington and London. 198 pages.

Krüger, L. 1913. Osmylidae. Beiträge zu einer Monographie der Neuropteren-Familie der Osmyliden, II. Stettiner Entomolgische Zeitung 74:3–123.

Krzeminski, W. 1991. Revision of the fossil Cylindrotomidae (Diptera, Nematocera) from Florissant and White River, USA. Palaeontologische-Zeitschrift 65 (3-4): 333–338.

Lakes, A. 1899. The Florissant basin: a prospecting trip in the volcanic region between Cripple Creek and South Park. Mines and Minerals 20:179–180.

Lambrecht, F. L. 1981. Tsetse flies and trypanosomiasis during the American Tertiary. National Geographic Society, Research Report 21:241–249.

LaMotte, R. S. 1952. Catalogue of Cenozoic plants of North America through 1950. Geological Society of America Memoir 51:1–381.

Leopold, E. B and S. T. Clay-Poole. 2001. Fossil leaf and pollen floras of Colorado compared: climatic implications. In Fossil flora and stratigraphy of the Florissant Formation, Colorado, edited by E. Evanoff, K. M. Gregory-Wodzicki, and K. R. Johnson. Proceedings of the Denver Museum of Nature & Science, ser. 4, 1:17–69.

Leopold, E. B., and H. D. MacGinitie. 1972. Development and affinities of Tertiary floras in the Rocky Mountains. In Floristics and paleofloristics of Asia and eastern North America, edited by A. Graham, 147–200. Elsevier, Amsterdam.

Lesquereux, L. 1873. Lignitic formation and fossil flora. Annual Report of the United States Geological Survey of the Territories 6:317–427.

Lesquereux, L. 1874. The lignitic formation and its fossil flora. Annual Report of the United States Geological Survey of the Territories 7:365–425.

Lesquereux, L. 1876. On the Tertiary flora of the North American lignitic, considered as evidence of the age of the formation. Annual Report of the United States Geological Survey of the Territories 8:275–315.

Lesquereux, L. 1878a. Remarks on specimens of Cretaceous and Tertiary plants secured by the survey in 1877; with a list of species hitherto described. Annual Report of the United States Geological Survey of the Territories 10:481–520.

Lesquereux, L. 1878b. Contributions to the fossil flora of the Western Territories, pt. II, The Tertiary flora. Report of the United States Geological Survey of the Territories 7:1–366, 65 plates.

Lesquereux, L. 1883. Contributions to the fossil flora of the Western Territories, pt. III, The Cretaceous and Tertiary floras. Report of the United States Geological Survey of the Territories 8:1–283, 60 plates.

Licht, E. L. 1987. Araneid fossils in volcanogenic and non-volcanogenic deposits: a comparison of information lost and found. Abstracts, Rocky Mountain Section, Geological Society of America, 40th Annual Meeting, Boulder, Colorado, 314.

Linsley, E. G. 1942. A review of the fossil Cerambycidae of North America. Proceedings of the New England Zoological Club 21:17–42.

Lundberg, J. G. 1975. The fossil catfishes of North America. Museum of Paleontology, University of Michigan, Papers on Paleontology 11 [Claude W. Hibbard Memorial Volume 2]: i–iv, 1–51.

MacGinitie, H. D. 1937. Stratigraphy and flora of the Florissant beds. Titles and abstracts of papers, Proceedings from the Paleontological Society, 362–363.

MacGinitie, H. D. 1953. Fossil plants of the Florissant beds, Colorado. Carnegie Institution of Washington Publication 599:1–198, plates 1–75.

MacGinitie, H. D. 1969. The Eocene Green River flora of northwestern Colorado and northeastern Utah. University of California Publications in Geological Sciences 83:1–203.

Manchester, S. R. 1987a. Extinct ulmaceous fruits from the Tertiary of Europe and western North America. Review of Paleobotany and Palynology 52:119–129.

Manchester, S. R. 1987b. The fossil history of the Juglandaceae. Missouri Botanical Garden Monographs in Systematic Botany 21:1–137.

Manchester, S. R. 1989. Attached reproductive and vegetative remains of the extinct American-European genus *Cedrelospermum* (Ulmaceae) from the early Tertiary of Utah and Colorado. American Journal of Botany 76 (2): 256–276.

Manchester, S. R. 1992. Flowers, fruits, and pollen of *Florissantia,* an extinct malvalean genus from the Eocene and Oligocene of western North America. American Journal of Botany 79 (9): 996–1008.

Manchester, S. R. 1999. Biogeographical relationships

of North American Tertiary floras. Annals of the Missouri Botanical Garden 86:472–522.

Manchester, S. R. 2001. Update on the megafossil flora of Florissant, Colorado. *In* Fossil flora and stratigraphy of the Florissant Formation, Colorado, edited by E. Evanoff, K. M. Gregory-Wodzicki, and K. R. Johnson. Proceedings of the Denver Museum of Nature & Science, ser. 4, 1:137–161.

Manchester, S. R., and P. R. Crane. 1983. Attached leaves, inflorescences, and fruits of *Fagopsis,* an extinct genus of fagaceous affinity from the Oligocene Florissant flora of Colorado, U. S. A. American Journal of Botany 70 (8): 1147–1164.

Manchester, S. R., and P. R. Crane. 1987. A new genus of Betulaceae from the Oligocene of western North America. Botanical Gazette 148 (2): 263–273.

Manchester, S. R., and M. J. Donoghue. 1995. Winged fruits of Linnaeeae (Caprifoliaceae) in the Tertiary of western North America: *Diplodipelta* gen. nov. International Journal of Plant Sciences 156 (5): 709–722.

Manwell, R. D. 1955. An insect Pompeii. The Scientific Monthly 80:356–361.

Mason, H. L. 1947. Evolution of certain floristic associations in western North America. Ecological Monographs 17:203–210.

Mason, W. R. M. 1976. A revision of *Dyscoletes* Haliday (Hymenoptera: Braconidae). Canadian Entomologist 108:855–858.

Mawdsley, J. 1992. A re-evaluation of the checkered beetles from the upper Miocene of Florissant, Colorado (Coleoptera: Cleridae). Psyche 99 (2-3): 129–139.

Mayor, A. G. 1924. Biographical memoir of Samuel Hubbard Scudder, 1837–1911. Memoirs of the National Academy of Sciences 17 (3): 79–104.

McLeroy, C. A., and R. Y. Anderson. 1966. Laminations of the Oligocene Florissant lake deposits, Colorado. Geological Society of America Bulletin 77:605–618.

Melander, A. L. 1946. Some fossil Diptera from Florissant, Colorado. Psyche 53 (3-4): 43–49, plate 2.

Melander, A. L. 1949. A report on some Miocene Diptera from Florissant, Colorado. American Museum Novitates 1407:1–63.

Menke, A. S., and A. P. Rasnitsyn 1987. Affinities of the fossil wasp, *Hoplisidea kohliana* Cockerell (Hymenoptera: Specidae: Specinae). Psyche 94 (1-2): 35–38.

Metcalf, Z. P. 1952. New names in the Homoptera. Journal of the Washington Academy of Sciences 42:226–231.

Meyer, H. W. 1986. An evaluation of the methods for estimating paleoaltitudes using Tertiary floras from the Rio Grande rift vicinity, New Mexico and Colorado. Ph.D. dissertation, University of California, Berkeley. i–vii + 217 pages.

Meyer, H. W. 1992. Lapse rates and other variables applied to estimating paleoaltitudes from fossil floras. Paleogeography, Paleoclimatology, Paleoecology 99:71–99.

Meyer, H. W. 1998. Development of an integrated paleontologic database to document the collections from Florissant, Colorado. Geological Society of America, Abstracts with Programs 30 (7): A263–A264.

Meyer, H. W. 2001. A review of the paleoelevation estimates for the Florissant flora, Colorado. *In* Fossil flora and stratigraphy of the Florissant Formation, Colorado, edited by E. Evanoff, K. M. Gregory-Wodzicki, and K. R. Johnson. Proceedings of the Denver Museum of Nature & Science, ser. 4, 1:205–216.

Meyer, H. W., L. Lutz-Ryan, M. S. Wasson, A. Cook, A. E. Kinchloe, and B. A. Drummond III. 2002. Research and educational applications of a web-based paleontologic database for Florissant Fossil Beds National Monument. Geological Society of America, Abstracts with Programs.

Meyer, H. W., and S. R. Manchester. 1997. The Oligocene Bridge Creek flora of the John Day Formation, Oregon. University of California Publications in Geological Sciences 141:i–xvi, 1–195, plates 1–75.

Meyer, H. W., and L. Weber. 1995. Florissant Fossil Beds National Monument: preservation of an ancient ecosystem. Rocks and Minerals 70:232–239.

Miller, J. Y., and F. M. Brown. 1989. A new Oligocene fossil butterfly, *Vanessa amerindica* (Lepidoptera: Nymphalidae), from the Florissant Formation, Colorado. Bulletin of the Allyn Museum 126:1–9.

Miner, R. W. 1926. A fossil Myriapod of the genus *Parajulus* from Florissant, Colorado. American

Museum Novitates 219:1–5.

Morse, A. P. 1911. The orthopterological work of Mr. S. H. Scudder, with personal reminiscences. Psyche 18 (6): 187–192.

Mosbrugger, V. 1999. The nearest living relative method. *In* Fossil plants and spores: modern techniques, edited by T. P. Jones and N. P. Rowe, 261–265. The Geological Society, London.

Navás, L. 1913. Neuropteros del R. Museo Zoológico de Napoles. Annuario del Museo Zoologico della Università di Napoli 4:1–11.

Nel, A., and J. Paicheler. 1994. Les Lestinoidia (Odoanta, Zygoptera) fossiles: un inventaire critique. Annales de Paleontologie 1, fasc. 1:1–2, 44–45, 53–56, 59.

Nieson, P. L. 1969. Stratigraphic relationships of the Florissant lake beds to the Thirtynine Mile volcanic field of central Colorado. Master's thesis, New Mexico Institute of Mining and Technology, Socorro. 65 pages.

O'Brien, N. R., and H. W. Meyer. 1996. The world of the micron at Florissant Fossil Beds National Monument. Park Science 16 (1): 22–23.

O'Brien, N. R., M. Fuller, and H. W. Meyer. 1998 Sedimentology and taphonomy of the Florissant lake shales, Colorado. Geological Society of America, Abstracts with Programs 30 (7): A162.

O'Brien, N. R., H. W. Meyer, K. Reilly, A. Ross, and S. Maguire. 2002. Microbial taphonomic processes in the fossilization of insects and plants in the late Eocene Florissant Formation, Colorado. Rocky Mountain Geology 37 (1): 1–11.

Olson, S. L. 1985. The fossil record of birds. *In* Avian biology, Vol. 8, edited by D. S. Farner, J. R. King, and K. C. Parkes, 79–252. Academic Press, New York.

Opler, P. A. 1982. Fossil leaf-mines of *Bucculatrix* (Lyonetiidae) on *Zelkova* (Ulmaceae) from Florissant, Colorado. Journal of the Lepidopterists' Society 36 (2): 145–147.

Osborn, H. F., W. B. Scott, and F. Speir Jr. 1878. Paleontological report of the Princeton Scientific Expedition of 1877. Museum of Geology and Archaeology of Princeton University Contributions 1:1–146, 10 plates.

Peale, A. C. 1874. Report of A. C. Peale M.D., geologist of the South Park division. Annual Report of the United States Geological and Geographical Survey of the Territories 7:193–273, 20 plates.

Pearl, R. M. 1953. A Colorado petrified forest. The Mineralogist 21 (4): 147–151.

Petrunkevitch, A. 1922. Tertiary spiders and opilionids of North America. Transactions of the Connecticut Academy of Arts and Sciences 25:211–279.

Poole, R. W., and P. Gentili, eds. 1996–1997. Nomina Insecta Nearctica: a check list of the insects of North America. Vol. 1, Coleoptera, Strepsiptera; Vol. 2, Hymenoptera, Mecoptera, Megaloptera, Neuroptera, Rhaphidioptera, Trichoptera; Vol. 3, Diptera, Lepidoptera, Siphonoptera; Vol. 4, Non-holometabolous orders. Entomological Information Services, Rockville, Md.

Popov, Yu. A. 1971. Historical development of Hemiptera of the Infraorder Nepomorpha (Heteroptera). Trudy Paleontologicheskogo Instituta, Akademiia Nauk SSSR 129:1–228, plates 1–9. [In Russian.]

Pulawski, W. J., and A. P. Rasnitsyn. 1980. On the taxonomic position of *Hoplisus sepultus* Cockerell, 1906, from the Lower Oligocene of Colorado (Hymenoptera, Sphecidae). Bulletin Entomologique de Pologne 50:393–396.

Rasnitsyn, A. P. 1986. Review of the fossil Tiphiidae with description of a new species (Hymenoptera). Psyche 93 (1-2): 91–101.

Rasnitsyn, A. P. 1993. Archeoscoliinae, an extinct subfamily of scoliid wasps (Insecta: Vespida = Hymenoptera: Scoliidae). Journal of Hymenoptera Research 2 (1): 85–96.

Rasnitsyn, A. P. 1995. Tertiary sawflies of the Tribe Xyelini (Insecta: Vespida = Hymenoptera: Xyelidae) and their relationship to the Mesozoic and modern faunas. Contributions in Science, Natural History Museum of Los Angeles County 450:1–14.

Rice, H. M. A. 1968. Two Tertiary sawflies (Hymenoptera: Tenthredinidae) from British Columbia. Geological Survey of Canada Paper 67-59:i–vi, 1–21.

Rodeck, H. G. 1938. Type specimens of fossils in the University of Colorado Museum. University of Colorado Studies 25 (4): 281–304.

Rohwer, S. A. 1908a. A fossil larrid wasp. Bulletin of the American Museum of Natural History

24:519–520

Rohwer, S. A. 1908b. A fossil mellinid wasp. Bulletin of the American Museum of Natural History 24:597 (art. 31).

Rohwer, S. A. 1908c. On the Tenthredinoidea of the Florissant Shales. Bulletin of the American Museum of Natural History 24:521–530 (art. 25).

Rohwer, S. A. 1908d. The Tertiary Tenthredinoidea of the expedition of 1908 to Florissant, Colorado. Bulletin of the American Museum of Natural History 24:591–595 (art. 30).

Rohwer, S. A. 1908e. Footnote to description of *Tenthredo saxorum* (Hymenoptera: Tenthredia). Bulletin of the American Museum of Natural History 24:594.

Rohwer, S. A. 1909a. The fossil Ceropalidae of Florissant, Colorado. Psyche 16:23–28.

Rohwer, S. A. 1909b. New Hymenoptera from western United States. Transactions of the American Entomological Society 35:99–136.

Rohwer, S. A. 1909c. Three new fossil insects from Florissant, Colorado. American Journal of Science, ser. 4, 27:533–536.

Rosen, D. E., and C. Patterson. 1969. The structure and relationships of the Paracanthopterygian fishes. Bulletin of the American Museum of Natural History 141:357–474, plates 52–78 (art. 3).

Schorn, H. E. 1998. *Holodiscus lisii* (Rosaceae): A new species of ocean spray from the late Eocene Florissant Formation, Colorado, USA. PaleoBios 18 (4): 21–24.

Scott, A. C., and F. R. Titchener. 1999. Techniques in the study of plant-arthropod interactions. *In* Fossil plants and spores: modern techniques, edited by T. P. Jones and N. P. Rowe, 310–315. The Geological Society, London.

Scott, G. R. 1975. Cenozoic surfaces and deposits in the southern Rocky Mountains. *In* Cenozoic history of the southern Rocky Mountains, edited by B. F. Curtis. Geological Society of America Memoir 144:227–248.

Scott, W. B. 1939. Some memoirs of a paleontologist. Princeton University Press, Princeton, N.J. 336 pages.

Scudder, S. H. 1876a. Fossil Coleoptera from the Rocky Mountain Tertiaries. Bulletin of the United States Geological and Geographical Survey of the Territories 2:77–87.

Scudder, S. H. 1876b. Critical and historical notes on forficulariae; including descriptions of new generic forms and an alphabetical synonymic list of the described species. Proceedings of the Boston Society of Natural History 18:287–332.

Scudder, S. H. 1876c (published in the Bulletin for 1875). Fossil Orthoptera from the Rocky Mountain Tertiaries. Bulletin of the United States Geological and Geographical Survey of the Territories 1, ser. 2, no. 6:447–449.

Scudder, S. H. 1878. An account of some insects of unusual interest from the Tertiary rocks of Colorado and Wyoming. Bulletin of the United States Geological and Geographical Survey of the Territories 4:519–543.

Scudder, S. H. 1880. The insect basin of Florissant. Psyche 3:77.

Scudder, S. H. 1881 (published in the Bulletin for 1882). Tertiary lake-basin at Florissant, Colorado, between South and Hayden Parks. Bulletin of the United States Geological and Geographical Survey of the Territories 6 (2): 279–300.

Scudder, S. H. 1882. Fossil spiders. Harvard University Bulletin 2 (9) (no. 22): 302–304.

Scudder, S. H. 1883a. Tertiary lake-basin at Florissant, Colorado, between South and Hayden Parks. Annual Report of the United States Geological and Geographical Survey of the Territories 12:271–293.

Scudder, S. H. 1883b. The fossil white ants of Colorado. Proceedings of the American Academy of Arts and Sciences 19:133–145.

Scudder, S. H. 1885a. Systematische Übersicht der fossilen Myriopoden, Arachnoiden, und Insekten. *In* Handbuch der Palaeontologie, Abt. I, Band 2, Mollusca und Arthropoda, edited by K. A. von Zittel, 720–833. R. Oldenbourg, Munich and Leipzig.

Scudder, S. H. 1885b. Description of an articulate of doubtful relationship from the Tertiary beds of Florissant, Colorado. Memoirs of the National Academy of Sciences 3 (6th memoir): 85–90.

Scudder, S. H. 1887. Fossil butterfly for sale. The Canadian Entomologist 19:120.

Scudder, S. H. 1889. The fossil butterflies of Florissant. United States Geological Survey Annual Report 8:439–474, plates 52–53.

Scudder, S. H. 1890a. The work of a decade upon fossil

insects; 1880–1890. Psyche 5 (165): 287–295.

Scudder, S. H. 1890b. The Tertiary insects of North America. Report of the United States Geological Survey of the Territories 13:1–734, 28 plates.

Scudder, S. H. 1890c. Physiognomy of the American Tertiary Hemiptera. Proceedings from the Boston Society of Natural History 24:562–579.

Scudder, S. H. 1890d. The fossil insect localities in the Rocky Mountain region. Psyche 5 (170): 363.

Scudder, S. H. 1892a. The Tertiary Rhynchophora of North America. Proceedings of the Boston Society of Natural History 25:370–386.

Scudder, S. H. 1892b. Some insects of special interest from Florissant, Colorado, and other points in the Tertiaries of Colorado and Utah. Bulletin of the United States Geological Survey 93:1–35, 3 plates.

Scudder, S. H. 1893. Tertiary Rhynchophorus Coleoptera of the United States. Monographs of the United States Geological Survey 21:1–206, 12 plates.

Scudder, S. H. 1894a. The American Tertiary Aphidae. Annual Report of the United States Geological Survey to the Secretary of the Interior 13:341–367.

Scudder, S. H. 1894b. Tertiary Tipulidae with special reference to those of Florissant Colorado. Proceedings of the American Philosophical Society 32:163–245, plates 1–9.

Scudder, S. H. 1900. Adephagous and clavicorn Coleoptera from the Tertiary deposits at Florissant, Colorado, with descriptions of a few other forms and a systematic list of the non-rhynchophorous Tertiary Coleoptera of North America. Monographs of the United States Geological Survey 40:1–148, 11 plates.

Sharov, A. G. 1962. Redescription of *Lithophotina floccosa* Cock. (Manteodea) with some notes on the manteod wing venation. Psyche 69 (3): 102–106.

Sharov, A. G. 1968. Phylogeny of the orthopteroid insects. Trudy Paleontologicheskogo Instituta, Akademiia Nauk SSSR 118:1–216 (see pp. 65–66), plates 1–12. [In Russian.] Translated by J. Salkind Israel Program for Scientific Translations, Jerusalem.

Shields, O. 1976. Fossil butterflies and the evolution of Lepidoptera. Journal of Research on the Lepidoptera 15 (3): 132–143.

Shufeldt, R. W. 1913. Fossil feathers and some heretofore undescribed fossil birds. Journal of Geology 21:628–652.

Shufeldt, R. W. 1917. Fossil remains of what appears to be a passerine bird from the Florissant shales of Colorado. Proceedings of the United States National Museum 53:453–455, plates 60–61.

Smith, D. M. 1998. Eating well in the Tertiary: insect herbivory in the Florissant fossil beds, Colorado. Geological Society of America, Abstracts with Programs 30 (7): A31–A32.

Smith, D. M. 2000a. The evolution of plant-insect interactions: insights from the fossil record. Ph.D. dissertation, University of Arizona, Tucson. 316 pages.

Smith, D. M. 2000b. Beetle taphonomy in a recent ephemeral lake, southeastern Arizona. Palaios 15:152–160.

Snyder, T. E. 1925. Notes on fossil termites with particular reference to Florissant, Colorado. Proceedings of the Biological Society of Washington 38:149–165, Chart 1.

Snyder, T. E. 1950. The fossil termites of the United States and their living relatives. Proceedings of the Entomological Society of Washington 52:190–193.

Steere, W. C. 1947. Cenozoic and Mesozoic bryophytes of North America. The American Midland Naturalist 36:298–324.

Størmer, L., A. Petrunkevitch, and J. W. Hedgpeth. 1955. Chelicerata with sections on Pycnogonida and Palaeoisopus *In* Arthropoda 2, pt. P of Treatise on invertebrate paleontology, edited by R. C. Moore. Geological Society of America, New York, and University of Kansas Press, Lawrence. 181 pages.

Štys, P., and P. Říha. 1977. An annotated catalogue of the fossil Alydidae (Heteroptera). Acta Universitatis Carolinae-Biologica 1974:173–188.

Tiffney B. H. 1981. Re-evaluation of *Geaster florissantensis* (Oligocene, North America). Transactions of the British Mycological Society 76 (3): 493–495.

Tindale, N. B. 1985. A butterfly-moth (Lepidoptera: Castniidae) from the Oligocene shales of Florissant, Colorado. Journal of Research on the Lepidoptera 24 (1): 31–40.

Ting, W. S. 1968. Fossil pollen grains of Coniferales from early Tertiary of Idaho, Nevada, and Colorado. Pollen et Spores 10 (1): 557–598.

Townes, H. 1966. Notes on three fossil genera of Ichneumonidae. Proceedings of the Entomological Society of Washington 68 (2): 132–135.

Trimble, D. E. 1980. Cenozoic tectonic history of the Great Plains contrasted with that of the southern Rocky Mountains: a synthesis. The Mountain Geologist 17 (3): 59–69.

Turner, R. E. 1912. Studies in the fossorial wasps of the family Scoliidae, subfamilies Elidinae and Anthoboscinae. Proceedings of the Zoological Society of London 46:696–754, plates 81–83.

Usinger, R. L. 1940. Fossil Lygaeidae (Hemiptera) from Florissant. Journal of Paleontology 14 (1): 79–80, plate 12.

Wang, Y., and S. R. Manchester. 2000. *Chaneya,* a new genus of winged fruit from the Tertiary of North America and eastern Asia. International Journal of Plant Sciences 161 (1): 167–178.

Warder, R. B. 1883. The silicified stumps of Colorado. Proceedings of the American Association for the Advancement of Science, Meeting Report 31:398–399.

Weber, W. A. 1965. Theodore Dru Alison Cockerell, 1866–1948. University of Colorado Studies, Series in Bibliography 1:1–124.

Weber, W. A., editor. 2000. The American Cockerell: a naturalist's life, 1866–1948. University Press of Colorado, Boulder. 351 pages.

Weilbacher, C. A. 1963. Interpretations of the laminations of the Oligocene Florissant lake deposits, Colorado. Ph.D. dissertation, University of New Mexico, Albuquerque. 115 pages.

Wetmore, A. 1925. The systematic position of *Palaeospiza bella* Allen, with observations on other fossil birds. Bulletin of the Museum of Comparative Zoology at Harvard College 67 (2): 183–193.

Wetmore, A. 1929. Birds of the past in North America. Smithsonian Institution Report for 1928, 377–389, plates 1–11.

Wheeler, E. A. 2001. Fossil dicotyledonous woods from Florissant Fossil Beds National Monument, Colorado. *In* Fossil flora and stratigraphy of the Florissant Formation, Colorado, edited by E. Evanoff, K. M. Gregory-Wodzicki, and K. R. Johnson. Proceedings of the Denver Museum of Nature & Science, ser. 4, 1:187–203.

Wheeler, W. M. 1906. The expedition to Colorado for fossil insects. American Museum Journal 6 (4): 199–203.

White, R. D. 1995. A type catalogue of fossil invertebrates (Arthropoda: Hexapoda) in the Yale Peabody Museum. Postilla 209:1–55.

Wickham, H. F. 1908. New fossil Elateridae from Florissant. American Journal of Science, ser. 4, 26:76–78.

Wickham, H. F. 1909. New fossil Coleoptera from Florissant. American Journal of Science, ser. 4, 28:126–130.

Wickham, H. F. 1910. New fossil Coleoptera from Florissant with notes on some already described. American Journal of Science, ser. 4, 29:47–51.

Wickham, H. F. 1911. Fossil Coleoptera from Florissant with descriptions of several new species. Bulletin of the American Museum of Natural History 30:53–69.

Wickham, H. F. 1912a. On some fossil rhynchophorous Coleoptera from Florissant, Colorado. Bulletin of the American Museum of Natural History 31:41–55, plates 1–4 (art. 4).

Wickham, H. F. 1912b. A report on some recent collections of fossil Coleoptera from the Miocene shales of Florissant. Bulletin of the Laboratories of Natural History of the State University of Iowa 6 (3): 3–38, plates 1–8.

Wickham, H. F. 1913a. Fossil Coleoptera from Florissant in the United States National Museum. United States National Museum Proceedings 45 (no. 1982): 283–303, plates 22–26.

Wickham, H. F. 1913b. Fossil Coleoptera from the Wilson ranch near Florissant, Colorado. Bulletin of the Laboratories of Natural History of the State University of Iowa 6 (4): 3–29, plates 1–7.

Wickham, H. F. 1913c. The Princeton collection of fossil beetles from Florissant. Annals of the Entomological Society of America 6:359–366, plates 38–41.

Wickham, H. F. 1914a. Twenty new Coleoptera from the Florissant shales. Transactions of the American Entomological Society 40:257–270, plates 5–8.

Wickham, H. F. 1914b. New Miocene Coleoptera from Florissant. Bulletin of the Museum of Comparative Zoology at Harvard College 58 (11): 423–494, plates 1–16.

Wickham, H. F. 1916a. The fossil Elateridae of

Florissant. Bulletin of the Museum of Comparative Zoology at Harvard College 60 (12): 493–527, plates 1–7.

Wickham, H. F. 1916b. New fossil Coleoptera from the Florissant beds. Bulletin of the Laboratories of Natural History of the State University of Iowa 7 (3): 3–20, plates 1–4.

Wickham, H. F. 1917. New species of fossil beetles from Florissant, Colorado. Proceedings of the United States National Museum 52 (no. 2189): 463–472, plates 37–39.

Wickham, H. F. 1920. A catalogue of the North American Coleoptera described as fossils. *In* Catalogue of the Coleoptera of America, north of Mexico, edited by C. W. Leng, 349–470. John D. Sherman Jr., Mount Vernon, N.Y.

Williston, S. W. 1886. Synopsis of North American Syrphidae. Bulletin of the United States National Museum 31:281–283.

Wilson, M. V. H. 1978. Paleogene insect faunas of western North America. Quaestiones Entomologicae 14:13–34.

Wing, S. L. 1987. Eocene and Oligocene floras and vegetation of the Rocky Mountains. Annals of the Missouri Botanical Garden 74:748–784.

Wingate, F. H., and D. J. Nichols. 2001. Palynology of the uppermost Eocene lacustrine deposits at Florissant Fossil Beds National Monument, Colorado. *In* Fossil flora and stratigraphy of the Florissant Formation, Colorado, edited by E. Evanoff, K. M. Gregory-Wodzicki, and K. R. Johnson. Proceedings of the Denver Museum of Nature & Science, ser. 4, 1:71–135.

Wobus, R. A., and R. C. Epis. 1978. Geologic map of the Florissant fifteen minute quadrangle, Park and Teller Counties, Colorado. United States Geological Survey Map I-1044.

Wodehouse, R. P. 1934. A Tertiary *Ephedra*. Torreya 34 (1): 1–4.

Wolfe, J. A. 1978. A paleobotanical interpretation of Tertiary climates in the Northern Hemisphere. American Scientist 66 (6): 694–703.

Wolfe, J. A. 1987a. Memorial to Harry D. MacGinitie (1896–1987). *In* Contributions to a symposium on the evolution of the modern flora of the northern Rocky Mountains. Annals of the Missouri Botanical Garden 74:684–688.

Wolfe, J. A. 1992. Climatic, floristic, and vegetational changes near the Eocene/Oligocene boundary in North America. *In* Eocene-Oligocene climatic and biotic evolution, edited by D. R. Prothero and W. A. Berggren, 421–436. Princeton University Press, Princeton, N.J.

Wolfe, J. A. 1994. Tertiary climatic changes at middle latitudes of western North America. Palaeogeography, Palaeoclimatology, Palaeoecology 108:195–205.

Wolfe, J. A. 1995. Paleoclimatic estimates from Tertiary leaf assemblages. Annual Review of Earth and Planetary Sciences 23:119–142.

Wolfe, J. A., C. E. Forest, and P. Molnar. 1998. Paleobotanical evidence of Eocene and Oligocene paleoaltitudes in midlatitude western North America. Geological Society of America Bulletin 110 (5): 664–678.

Wolfe, J. A., and H. E. Schorn. 1990. Taxonomic revision of the Spermatopsida of the Oligocene Creede flora, southern Colorado. U.S. Geological Survey Bulletin 1923:i–iv, 1–40, plates 1–13.

Wolfe, J. A., and R. A. Spicer. 1999. Fossil leaf character states: multivariate analyses. *In* Fossil plants and spores: modern techniques, edited by T. P. Jones and N. P. Rowe, 233–239. The Geological Society, London.

Wolfe, J. A., and T. Tanai. 1987. Systematics, phylogeny, and distribution of *Acer* (maples) in the Cenozoic of western North America. Journal of the Faculty of Science, Hokkaido University, ser. 4, 22:1–246.

Wolfe, J. A., and W. Wehr. 1987. Middle Eocene dicotyledonous plants from Republic, northeastern Washington. U.S. Geological Survey Bulletin 1597:i–iv, 1–25, plates 1–16.

Zeuner, F. E. 1941. The fossil Acrididae (Orth. Salt.), pt. 1, Catantopinae. Annals and Magazine of Natural History, ser. 11, 8:510–522.

Zeuner, F. E., and F. J. Manning. 1976. A monograph on fossil bees (Hymenoptera: Apoidea). Bulletin of the British Museum (Natural History), Geology 27 (3): 149–268, plates 1–4.

Zhelochovtzev, A. N., and A. P. Rasnitsyn. 1972. On some Tertiary sawflies (Hymenoptera: Symphyta) from Colorado. Psyche 79 (4): 315–327.

INDEX

Page numbers in **boldface** indicate figures. Taxa from Appendix 1 are listed for all families and selected genera of the plants, and for all orders and all families of the insects.

Abies: habitat, 54; pollen, **121,** 122, 191; seed, **87,** 87, 199
Acari, 133, 202
Acer, 47, 54, **109,** 110–111, 194, 200
Aceraceae, 194, 200
Achilidae, 207
Acrididae, **139,** 203
Aeshnidae, 203
Agaonidae, 228
age: of Florissant fossils, 42–43
Agromyzidae, 223
Agyritidae, 209
Ailanthus, 47, **110,** 111, 194
Alydidae, 206
Amelanchier, 54, 102, **103,** 193
American Museum of Natural History, 11, 233
Amia, **176–177,** 176–179, 231
Amiidae, **176–177,** 176–179, 231
Amyzon, **177,** 179, 231
Anacardiaceae, 111–113, **112,** 194, 200
Andrenidae, 229
angiosperms: leaves, fruits, and seeds, 89–117, 192–195; pollen 196–201; woods, 79–80, **79–80,** 191
Anobiidae, 213
Anthicidae, 215

Anthomyiidae, **160,** 223
Anthophoridae, 229
Anthribidae, 216
ants. *See* Hymenoptera
aphids. *See* Homoptera
Aphredoderidae, **178,** 179, 231
Apidae, 229
Apocynaceae, 195, 201
apples. See *Malus*
Arachnida, 130–133, 201
Araliaceae, **114,** 115, 194
Araneae, **130–132,** 131–133, 201
Araneidae, **131–132,** 201
Arecaceae, 117, 195, 201. *See also* palms
Argidae, 225
Aristolochiaceae, 192
ash, volcanic, 29–31, **32, 33**
Asilidae, 157, **159,** 221
Aspleniaceae, 191
Asteraceae, 120–121, 197
Asterocarpinus, 94–97, **95,** 192
Attelabidae, **155,** 216
Aulacidae, 226

bees. *See* Hymenoptera
beetles. *See* Coleoptera
Berberidaceae, **90,** 90–91, 192, 199
Bethylidae, 228
Betulaceae, 94–97, **95,** 192, 199
Bibionidae, **158,** 220
Big Stump, 14, **75,** 77; attempted removal of, 6–7, **7;** location relative to Scudder's site, **8**
biogeographic dispersal, 74; during Eocene-Oligocene, 59–61, **60;** of particular taxa, 97, 99, 101, **161**
biotic community, 46–48; changes during Eocene-Oligocene, 58–61, **60;** distribution of modern relatives, 47–48; evolution and mechanisms of change, 48, **60,** 61; microhabitats, 53–54; origin of the Eocene Florissant community, 57
birds, 180–183, 231
Bivalvia, **171,** 172, 231
Blasticotomidae, 225
Blattaria, 141, **142,** 204
Blattelidae, 204
Bombacaceae, 122
Bombyliidae, 157, **159,** 222
Bostrichidae, 213
bowfin. *See* Amiidae
Brachypteridae, 213
Braconidae, **167,** 226
Brentidae, 216
Bridge Creek flora, 59, **60**
British Museum, 11; postcard specimens, **85, 112.** *See also* Natural History Museum, London
brontotheres, 42, **183,** 184–185; extinction of, 61, 185
Brontotheriidae, **183,** 184–185, 232
bugs. *See* Hemiptera
Buprestidae, **152,** 211
Burseraceae, 111, 194
butterflies. *See* Lepidoptera
Buxaceae, 122
Byrrhidae, 211

caddisflies. *See* Trichoptera
Calopterygidae, 203
Cannabaceae, 102, **102**, 193
Cantharidae, **153**, 212
Caprifoliaceae, 195, 200
Carabidae, **150**, 208
Carya, 54, **96–97**, 97, 192, 200
Caryophyllaceae, 199
Castniidae, 224
catfish. *See* Ictaluridae
Catostomidae, 177, 179, 231
cattail. *See* Typhaceae
Cecidomyiidae, 221
Cedrela, 111, **111**, 194
Cedrelospermum, 45; habitat, 54, 101; leaves and fruits, 99–101, **100**, 193; past distribution, 47, 101; possible wood of, 79–80
Celastraceae, 194
Cephidae, 226
Cerambycidae, **154**, 215
Cercidiphyllaceae, 122
Cercocarpus, 47, 54, 102, **103**, 193
Cercopidae, **147**, 207
Chadronoxylon, 80, **80**, 191
Chaeteessidae, 204
Chalcididae, **167**, 228
Chamaecyparis, 54, **84**, 84–85, 191
Charadriiformes, **182**, 231
Chelonariidae, 211
Chenopodiaceae/Amaranthaceae, 54, 119, **120**, 122, 199
Chironomidae, 157, **158**, 221
Chrysididae, 228
Chrysomelidae, **155**, 215
Chrysopidae, **149**, 208
Cicadellidae, 207
Cicadidae, **147**, 207
Cimbicidae, 225
Cixiidae, 207
clams, **171**, 172. *See also* Bivalvia
Cleridae, 213
climate. *See* paleoclimate
Clubionidae, 202
Coccidae, 208
Coccinellidae, **153**, 214
Cockerell, T. D. A., 11, **11, 112, 159**
cockroaches. *See* Blattaria
Coenagrionidae, 203
Coleoptera; 148, 150, **150–156**, 208

Colorado Midland Railway, 5, **6**, 14, **15**
Colorado Petrified Forest, 14, **15**
Colydiidae, 214
community. *See* biotic community
conifers: leaves, seeds, and cones, 83–89; pollen, 120, 122; taxonomic lists, 191, 196, 199; wood, 77–79
conservation: of the fossil beds, 14–19
Convolvulaceae, 195
Cope, E. D., 13, 176, 235
Coraciiformes, **180**, 181, 231
Coreidae, 145, 206
Corixidae, 145, 204
Cossidae, 224
Crataegus, 102, **103**, 193
Creede flora, 59–61, **60**, 87
Cryptophagidae, 213
cuckoo. *See* Cuculidae
Cucujidae, 213
Cuculidae, 180, **180**, 231
Cupressaceae: leaves and cones, **84**, 84–85, 87, 191; pollen record, 56, 119, 199
Curculionidae, **156**, 217
cuticle, **81**
Cydnidae, 145, 206
Cynipidae, 228
Cyperaceae, 195

Dascillidae, 210
dates, radiometric, 42
Dermaptera, 143, **143**, 204
Dermestidae, 212
Diapriidae, 228
diatoms, 123–125, **124**; in preservation of the fossils, **35**, 35–37, **36**; sedimentation of, 29–31, **32, 33, 34**
Dictyopharidae, **147**, 207
Didelphidae, **181**, 183, 231
Dioscoreaceae, 116, 195
Diptera, 157–158, **157–161**, 219
Disney, W., 15–16, **17**
dispersal. *See* biogeographic dispersal
dragonflies. *See* Odonata
Drepanosiphidae, **148**, 208
Dryopidae, 211

Dytiscidae, **151**, 209

earwigs. *See* Dermaptera
ecosystem. *See* biotic community
Elaeagnaceae, 122, 200
Elateridae, **153**, 211
elevation. *See* paleoelevation
elm family. *See* Ulmaceae
Embiidae, 204
Embiidina, **143**, 143–144, 204
Empididae, 157, **159**, 222
Eocene: boundary of, 42; climate, 56
Eocene-Oligocene transition, 58–61, **60**
Eomeropidae, 219
Ephedra, 54, 122, 191, 199
Ephedraceae, 191, 199
Ephemeridae, 202
Ephemeroptera, 135–136, **136**, 202
Equidae, 184, **184**, 231
Equisetaceae, **82**, 191
Ericaceae, 122, 200
Eriophyidae, 202
Erotylidae, 214
Eucnemidae, 212
Eucommia, 47, **118**, 118–119, 122, 194, 200
Eucommiaceae, 194, 200
Eumasticidae, 203
Euphorbiaceae, 193, 200
Eurytomidae, 228
Fabaceae, 102, **105–106**, 106, 191, 193, 200
Fagaceae, 80, **90–92**, 91–94, **94**, 119, 192, 199
Fagopsis: cuticle, **81**; extinction, 61; habitat, 54; leaves and reproductive organs, **90–92**, 91–93; pollen, **91**; possible wood of, 80; reconstruction of, **92**; taxonomic list, 192, 199
ferns, 56, 83, 120, 191, 199
Figitidae, 228
fir. *See Abies*
fish, 176–179, 231
flies. *See* Diptera
Florissant: establishment of town, 5; map of area, **2**. *See also* Lake Florissant
Florissant Formation, 38–41;

geologic map, **30**; stratigraphy, **39**
Florissant Fossil Beds National Monument, **1–2**, **18**, 18–19; collections, 234; location, **2**; wasp logo, see *Palaeovespa*
Florissant Fossil Quarry, 16–17; **33**
Florissantia: insect, **147**, 207; plant, 97–99, **99**, 193, 200
flowering plants. *See* angiosperms
form genera, **105**, 106, 113
Formicidae, **170**, 230
fossilization: processes of, 27–28, 34–37, **35–37**, 77; taphonomic biases, 69–70, 175
Fulgoridae, 207

Gastropoda, **171**, 172–173, 231
geologic map, **30**
geologic time chart, **22**
Geometridae, 225
Geophila, 231
Gerridae, 145, **145**, 205
Glossina, 47, 223
Gnaphosidae, **131**, 201
golden-rain tree. *See Koelreuteria*
Gramineae. *See* Poaceae
grasses, 122, 140. *See also* Poaceae
grasshoppers. *See* Orthoptera
Green River flora, 57
Grimmiaceae, 191
Grossulariaceae, 102, 193
Gryllacrididae, 204
Gryllidae, **140**, 204
Guffey volcano, **25**, 25–29, **26, 28**, 37, 38

Haglidae, 204
Halictidae, 229
Harvard University, Museum of Comparative Zoology, 10, 234
harvestmen. *See* Opiliones
hawthorn. *See Crataegus*
Hayden Survey, 7, 8, 10
Heleomyzidae, 223
Hemiptera, 144–145, **144–146**, 204

hickory. See *Carya*
Hodotermitidae, 204
Homoeogamiidae, 204
Homoptera, 146–148, **147–148**, 207
hoppers. See Homoptera
horses. See Equidae
horsetails, **82**, 83
Humulus, 102, **102**, 193
Hydrangea, 47, 113–115, **114**, 194
Hydrangeaceae, 113–115, **114**, 194
Hydrophilidae, 210
Hydropsyche, 47, **161**, 224
Hydropsychidae, **161**, 223
Hymenoptera, 165–166, **166–170**, 225
Hyracodontidae, 232

Ibaliidae, 228
Ichneumonidae, **167**, 226
Ictaluridae, **178**, 179, 231
Inocelliidae, 208
insects, 134–170; metamorphosis, 135; taxonomic lists, 202–230. See also plant-insect interactions
Isoptera, 141, **142**, 204

Juglandaceae, **96–97**, 97, 119, 192, 200
Julida, 202
Julidae, 202

Kalotermitidae, 204
Koelreuteria: habitat, 54; leaves and fruits, 106, **108**, 193; present distribution, 47; wood, 79, **79**, 191

Labiduridae, **143**, 204
lacewings. See Neuroptera
lahars, **25**, 25–29, **26, 28, 30**, 77
Lake Florissant, duration of existence, 41; formation of, **25, 28**, 29; sedimentation in, 29–34, **32–34**, 38–41; size and shape of, 29, **30**; water conditions in, 123–125, 171, 176
Lakes, A., 8; geologic map, **9**
Lampyridae, 212
Laramide Orogeny, 22–23

Latridiidae, 214
Lauraceae, 89–90, **90**, 192
Lauxaniidae, 223
Leguminosae. See Fabaceae
Leiobunidae, **134**, 202
Leiodidae, 209
Lemnaceae, 195
Leopold, E., 17, 119
Lepidoptera, 163–165, **163–165**, 224
Leptoceridae, 224
Leptomerycidae, 232
Leptophlebiidae, 202
Lesquereux, L., 10, **10**
Libytheidae, 224
Limnephilidae, 224
Linyphiidae, 201
living fossil, 178
Lucanidae, 210
Lycidae, 212
Lycosidae, **132**, 202
Lygaeidae, 145, **146**, 205
Lymexylidae, 213
Lymnophila, 231
Lyonetiidae, 224

MacGinitie, H. D., 3, 12, **12**, 17, **76**
Mahonia, **90**, 90–91, 192, 199
Malus, 102, **104**, 193
Malvaceae, 122, 200
mammoth, 42
Mantidae, 204
mantids. See Mantodea
Mantodea, 141, **141**, 204
maples. See *Acer*
mayflies. See Ephemeroptera
Mecoptera, **156**, 156–157, 219
Megachilidae, **168**, 229
Megapodagrionidae, 203
Meliaceae, 111, **111**, 194
Melittidae, 229
Meloidae, 215
Melyridae, 213
Merycoidodontidae, **183**, 185, 232
Mesohippus, 184, **184**, 231
millipedes, 134, **135**, 202
Mindaridae, 208
Miridae, 145, **146**, 205
mites and ticks. See Acari
mollusks, 172–173. See also Bivalvia; Gastropoda
Moraceae, 101, **101**, 193, 200
Mordellidae, **154**, 214
mosses, 81–83, **82**, 191

multiple-organ reconstruction, 74, **92**, 93, 101
Muscidae, 157, **161**, 223
Mycetophagidae, 214
Mycetophilidae, 220
Mydaidae, 221
Myrtaceae, 193

Natural History Museum, London, 11, 234
Nemestrinidae, 222
Nemopteridae, **149**, 208
Nepidae, **144**, 145, 204
Neuroptera, 148, **149**, 208
Nitidulidae, 213
Notonectidae, **144**, 145, 204
Nymphaeaceae, **120**, 121, 199
Nymphalidae, **164**, 224

oaks. See *Quercus*
Odonata, 136–138, **137–138**, 203
Odontoceridae, 224
Oecophoridae, 224
Oedemeridae, 215
Oleaceae, 195, 201
Onagraceae, **121**, 122, 193, 200
Opiliones, 133, **134**, 202
opossum. See Didelphidae
orb-weaving spiders. See Araneidae
oreodonts. See Merycoidodontidae
Orthoptera, 138–140, **139–140**, 203
Osmylidae, 208
Ostracoda, **162**, 171, **171**, 230
Otitidae, 223

Palaeovespa, **169**, 230
paleoclimate, 48–53; growth rings in trees, 52–53, 77, 79; nearest living relative method, 49–50; physiognomic method, 50–51; pollen and spore evidence, 52; precipitation, 52; temperature, 51–52
paleoecosystem. See biotic community
paleoelevation, 62–63
Palmae. See Arecaceae
palms, 50, 122. See also Arecaceae

Pamphiliidae, 225
Panorpidae, **156**, 219
paper shale, 31–34, **32, 33**
Paracarpinus: habitat, 54; insect damage, **65**; leaves, 94–97, **95**, 192
Parajulidae, 202
Parattidae, 202
Peale, A. C., 7, 75
Pentatomidae, 145, 206
petrified forest, 5–7, 14–16, **75–76**, 75–80; process of fossilization, 27–28, 77; size of trees, **75**, 77
Phasmatidae, 203
Phasmida, 203
Phoridae, 223
Phrygancidae, 224
Pieridae, **163**, 224
Piesmatidae, 205
Pike Petrified Forest, 14–15, **16**
Pikes Peak Granite, 23, **23**
Pinaceae, leaves and seeds, 87–89, **87–89**, 191; pollen record, 56, 119, 199
pine. See *Pinus*
Pinus, 47, 54, 87–89, **88–89**, 191–192, 199
Piophilidae, 223
Pipunculidae, 223
pirate perches. See Aphredoderidae
plant-insect interactions, 63–69, **64–67**, 144–145, 148, 165
Platanaceae, 192
Platypezidae, 223
Pleistocene deposits, 41–42
Poaceae, **116**, 116–117, 195, 201. See also grasses
Polemoniaceae, 122
pollen and spores, 117–123; as evidence for paleoclimate, 52; as evidence for vegetation change, 54–56; comparison with leaf record, 119, 122; dominant families, 119; naming, 118–119; pollen diagram, **55**; taphonomy, 119; taxa known exclusively from, 120–122; taxonomic lists, 196–201

Polycentropidae, 224
Polygonaceae, 122
Polypodiaceae, 199
Polystoechotidae, 208
Pompilidae, **168,** 229
Populus, 54, 97, **98,** 192, 200
Potamogetonaceae, 195
preservation. *See* conservation; fossilization
Princeton Expedition of 1877, 10–11; collections of, 234, 235
Proctotrupidae, 228
Prodryas, 46, 163–165, **164,** 225
Psephenidae, 211
Pseudolestidae, 203
Psychomyiidae, 224
Psyllidae, 208
Pteromalidae, 228
Ptychopteridae, 221
Pyrochroidae, 215
pyroclastic flow, 24
Pyrrhocoridae, 206

Quercus, 54, 93–94, **94,** 192, 199

Raphidiidae, **149,** 208
Reduviidae, 145, **146,** 205
redwoods. *See Sequoia*
Redwood Trio, 14, **76,** 77
Rhagionidae, 221
Rhamnaceae, 113, **113,** 194
Rhinotermitidae, 204
Rhipiphoridae, 214
Rhoipteleaceae, 121, 199
Rhus, 54, **112,** 113, 194
Richardiidae, 223
rollers. *See* Coraciiformes
Rosa, 47, 54, 102, **104,** 193
Rosaceae, 102, **103–104,** 193, 200
rose family. *See* Rosaceae
Rutaceae, **110,** 111, 194, 200

Salicaceae, 97, **98,** 192, 200
Salix, 47, 54, 97, **98,** 192, 200
Salpingidae, 215
Sambucaceae, 115, **115,** 195, 200
Sapindaceae, 106, **108–109,** 191, 193, 200
Sapotaceae, 122
Saturniidae, 225
Scarabaeidae, **152,** 210
Scathophagidae, 223
Scatopsidae, 221
Scelionidae, 228
Schizaeaceae, 199
Sciaridae, 220
Sciomyzidae, 223
Scirtidae, 210
Scoliidae, **169,** 230
Scolytidae, 219
scorpionflies. *See* Mecoptera
Scudder, S. H., 7–10, **8**
Segestriidae, **130,** 201
Selaginellaceae, 199
Sepsidae, 223
Sequoia: differences between fossil and modern, 77, 86, 123; habitat, 54; leaves and cones, **85–86,** 85–87, 191; past and present distribution, 47; pollen, **122,** 123, 196, 199; trees and wood, **75, 76,** 77, **78,** 191
Sequoiaoxylon, 77, 191
serviceberry. *See Amelanchier*
shale. *See* paper shale
Silphidae, 209
Simaroubaceae, **110,** 111, 194
Siphlonuridae, 202
Siricidae, 226
Smilacaceae, 116, 195
Smithsonian, National Museum of Natural History, 10, 11, 235
snails, **171,** 172–173. *See also* Gastropoda

snakeflies. *See* Neuroptera
Solanaceae, 121, 201
Sparganiaceae, 201
Sphecidae, 228
spiders. *See* Araneae
Staphylea, 54, 68, 106, **107,** 193
Staphyleaceae, 106, **107,** 193
Staphylinidae, **151,** 209
Stephanidae, 228
Sterculiaceae, 97–99, 193, 200
Stratiomyidae, 221
Styracaceae, 192
suckers. *See* Catostomidae
Synchroidae, 214
synonymy, 12, 127–128
Syrphidae, 157, **160,** 223

Tabanidae, 221
taphonomy, 12, 69–70, 80, 119, 175
Taxaceae, 83, **84,** 191, 199
Taxodiaceae, **85–86,** 85–87, 119, 191, 199
Tenebrionidae, **154,** 214
Tenthredinidae, **166,** 225
termites. *See* Isoptera
Tetragnathidae, 202
Tettigoniidae, **139,** 203
Theraphosidae, **130,** 201
Therevidae, 221
Thomisidae, 202
Thouinia, 47, 106, **109,** 106, 117, 194
Throscidae, 212
Thymelaeaceae, 193, 200
Tiliaceae, 192
Tingidae, 145, **145,** 205
Tiphiidae, 229
Tipulidae, 157, **157,** 219
Torreya, 83, **84,** 191
Tortricidae, **163,** 224
Torymidae, 228
tree of heaven. *See Ailanthus*
Trichophanes, **178,** 179, 231

Trichoptera, 158, 160, **161–162,** 163, 223
Trogidae, 210
Trogossitidae, 213
tsetse flies, 50. See also *Glossina*
tuff, 31, **32–33,** 34
Typha, 53, **116,** 117, 201
Typhaceae, **116,** 117, 201

Ulmaceae, 99–101, **100,** 119, 191, 193, 200
University of California, Museum of Paleontology, 12, 235

Vanessa, 47, **164,** 165, 225
Veliidae, 145, 204
Vespidae, **169,** 230
Vitaceae, 194, 200
volcanic eruptions: effects on vegetation, 54–56. *See also* ash, volcanic; pyroclastic flow

walking sticks. *See* Phasmida
Wall Mountain Tuff, **24,** 24–25
walnut family. *See* Juglandaceae
wasps. *See* Hymenoptera
web-spinners. *See* Embiidina
wood, 77–80, **78–80**

Xyelidae, 225
Xylomyidae, 221
Xylophagidae, 221

Yale University, Peabody Museum, 11, 235

Zetobora, **142,** 204